ENVIRONMENTAL SCIENCE, ENGINEERING AND TECHNOLOGY

ARID AND SEMI-ARID ENVIRONMENTS

BIOGEODIVERSITY, IMPACTS AND ENVIRONMENTAL CHALLENGES

Environmental Science, Engineering and Technology

Additional books in this series can be found on Nova's website under the Series tab.

Additional e-books in this series can be found on Nova's website under the e-book tab.

ENVIRONMENTAL SCIENCE, ENGINEERING AND TECHNOLOGY

ARID AND SEMI-ARID ENVIRONMENTS

BIOGEODIVERSITY, IMPACTS AND ENVIRONMENTAL CHALLENGES

GABRIEL FERGUSON
EDITOR

nova publishers
New York

Library of Congress Cataloging-in-Publication Data

ISBN: 978-1-63483-383-7

Published by Nova Science Publishers, Inc. † New York

CONTENTS

PREFACE

Water is the most limiting factor for irrigated agriculture in arid and semi-arid areas of European Mediterranean countries. In this book, the authors' explore the different mechanisms and robust tools to monitor plant-water status, with the aim of keeping crops within a certain threshold of moderate-to-mild water stress. Other chapters include research on agricultural techniques in semi-arid environments that would benefit the surrounding environment and impact soil management. The third chapter includes site-specific documentation of landforms developed in the Ejina Basin in Central Asia and its implications for late quaternary landscape evolution and palaeoenvironmental change. The fourth chapter focuses on the links between economic value addition, demographics, personal income and entrepreneurship in selected South African towns. The last chapter reviews thirty years of ecological monitoring in Algerian arid rangelands.

Chapter 1 – The soil erosion, runoff and nutrient-loss patterns were monitored in a mountainside with olive (*Olea europaea* cv. Picual) plantations under conventional and conservation agricultural systems based in two different soil management: 1) conventional tillage and 2) non-tillage with barley (*Hordeum vulgare*) strips of 4 m width, respectively. The erosion plots, located in Lanjaron on the southern flank of the Sierra Nevada Mountains in SE Spain, had 30% slope at an altitude of 565 m and 192 m^2 (24 m x 8 m) in area. The highest erosion rate of 3.5 t ha^{-1} was registered under conventional system and runoff of 11.4 mm under conservation system, over the entire study period (19 erosive events). The conservation system reduced the soil erosion by 69% with respect to the conventional system, and the runoff under conventional system in relation to conservation system by 54%. The total NPK fluxes by runoff from conventional system amounted to 2.87, 0.09 and 0.93 kg ha^{-1}, and from conservation system 1.57, 0.09 and 0.98 kg ha^{-1}, respectively. In general, the soluble average concentrations of plant nutrients (N-NO_3, N-NH_4, P-H_2PO_4, and K) in the surface runoff were higher than the recommended level for standard water quality for from conventional system. These findings support the recommendation of conservation agricultural techniques for semiarid and sloped agricultural land with rainfed-tree crops for erosion and pollution control.

Chapter 2 – Water is the most limiting factor for irrigated agriculture in arid and semi-arid areas of European Mediterranean countries. The searching of alternative irrigation strategies such as deficit irrigation and the use of different tools to monitor the crop-water status are some options to improve the water management under limiting drought conditions. Infrared thermography is a non-invasive promising technique to monitor the crop-water status based on the relations between the leaf or canopy temperature and the transpiration rate, which is largely related with the stomatal conductance. From an energy balance point of view, as a consequence of stomatal regulation (partial closure) under a mild or moderate water stress, leaf temperature tends to increase because of a decrease in the heat dissipation associated with the transpiration process, which is known as the evaporative cooling process. However, temperature readings are highly dependent on many meteorological variables being very difficult to define threshold values of canopy temperature for the irrigation scheduling or the crop-water status monitoring. For this reason, in spite of many works devoted to the use of infrared thermography, there are many questions and doubts that have to be solved for real optimization, protocolling and standardization of this technique, (such as, when the images should be taken, the best time along the day, the most representative, and relevant water stress index or the image processing). Taking into account these considerations, the aim of this work was to respond to some questions that could help to optimize the use of this technique in some woody crops: i) which are the main advantages and disadvantages of this technique?; ii) which of the actual thermal indices is the most useful and representative?; iii) when is the best moment to assess the crop physiological status using the canopy temperature, focusing the answers in woody crops.

Chapter 3 – The north-south trending basin of Ejina in Inner Mongolia of northernwestern China, Central Asia is located at the lower reaches of the Heihe River watersheds and is a regional tectonic depression limited to the west part of the Alashan Plateau. Though neotectonic activity is believed to be the major factor in the evolution of the topography, detailed documentation and analysis of climatic landscape features like alluvial fans, desert plains, aeolian dunes and bedrock gorges and their environmental implications are lacking. The present study is a site-specific documentation of landforms developed in the wide part of the Ejina Basin, with the aim to identify the climatic landforms based on the method of climatic geomorphology and to evaluate its landscape evolution and response to palaeoclimate changes. The morphodynamics of older landscapes are recognized by making comparison with the present climate and its corresponding landscapes. Occurrence of desert plains and aeolian dunes in the central basin is inferred to be the products of modern hyper-arid climate. No active geomorphoclimatic dynamics for development of desert bedrock gorges and pediments is observed at the margins in the present day conditions, thus we infer an obvious linkage of these landforms to the humid phases, which provided high energy runoff for the formation of landforms associated erosional features. Clear evidences testifying the basin-scale shifting and transformation of different morphoclimatic zones in the Ejina basin are observed, which prove that the main geomorphic unit is changed from an alluvial-lacustrine plain to a desert plain. Pediments and desert gorges are both relict landforms associated with more humid and colder climate during the last glacial period. However, the coexistence of diverse landscapes and the consequent geomorphodiversity in the basin should be a compound result of surficial processes other than glaciations. Fluvial/alluvial actions, aeolian, abrasion/deflation, aggradation and degradation processes together with climatic changes have conjunctionally played important role. The climate and hydrological conditions of the basin during the last glaciation and during the Early Holocene were much better than at present, possibly having an average annual precipitation ranged between 60~350 mm on the margins of the basin and being characterized in the central basin by high lake levels during ca. 39-23 ka BP but great fluctuations during Holocene. Because the Asian summer monsoon was quite weak during the glacial times, the periods of lower aridity during the late Pleistocene in the Ejina Bain could be induced by an increase of the westerlies and a weakening of the Asian winter monsoon on the arid areas of the central Asia.

Chapter 4 – The landscape of the Karoo, the arid to semi-arid heartland of South Africa, has for millennia been subjected to a conservation/utilization dilemma. At first hunter-gatherer San tribes lived off Karoo plants and animals. Some two thousand years ago Khoikhoi herders and their livestock arrived. In the 18th Century European farmer settlers moved into the Karoo and replaced the San and Khoikhoi. In time they became wool producers and suppliers to Industrial Britain. Wool was a major export staple of the Cape Colony. After World War II wool passed its export pinnacle and by the 1960s some Karoo towns were regressing. Increased poverty was a big concern and numerous government-instigated investigations followed. Since then some Karoo towns have prospered in post-productive fashion. Now the Karoo faces new development challenges. South Africa's part of the Square Kilometre Array, a large radio telescope, is being developed in the western Great Karoo. Some large solar power installations have also been erected. The Karoo was also identified as a likely source of shale gas. The South African government recently decided to allow exploration to proceed, an action that has occasioned vociferous opposition and rekindled the conservation/utilization dilemma. Exploration could last a decade and the period should be used to gather facts needed for wise and transparent decisions. This chapter contributes to this by focusing on the links between economic value addition, demographics, personal income and entrepreneurship in selected Karoo towns. Statistical and other techniques are used to quantify relationships between gross economic value addition, town populations, total personal income and total number of enterprises in Karoo towns. A power law describes the enterprise richness of Karoo towns and an analysis of functional business diversity indicates a tipping point at about 100 enterprises. The ways by which entrepreneurial spaces are used in Karoo towns are elucidated and a predictive capability is developed about the likely responses of populations, entrepreneurs and employment in Karoo towns to economic impacts, be they from the exploitation of shale gas or other developments. A lack of local information forced the use of US information about drilling rig numbers and densities in shale gas and oil extraction. This enables predictions indicating substantial economic and other benefits to some Karoo towns from shale gas extraction. These benefits have to be weighed against potential detrimental impacts of gas extraction in order to seek possible win-win solutions.

Chapter 5 – In 1978, several observatories, aiming ecological monitoring, were settled in the arid high plains of South west Algeria. These observatories are now integrated in an international network called ROSELT (Long Term Ecological Monitoring Observatories Network) administered by the OSS (Observatory of the Sahara and the Sahel). A study area included in one of these observatories was monitored on more than thirty years. It is part of Algerian steppe, the most widespread of the Maghreb. The results showed that this area has undergone profound changes.

These modifications concerned both physical and biological components. The parameters considered include vegetation, flora, and soil surface properties. It appears that land cover has changed and former vegetation dominated by *Stipa tenacissima, Lygeum spartum* and *Artemisia herba alba* has disappeared. New species, like *Atractyis serratuloides, Salsola vermiculata* and *Noaea mucronata*, rare in 1978 are now dominant, in relation with a dynamic of degradation. The largest decline is observed for *Stipa tenacissima* vegetation units constituting 2/3 of the landscape in 1978 and occupying just 1/10 in 2010. The vegetation cover has also significantly declined. In 1978, the average coverage was over 34%, whereas now it is less than 19% of the total. The biodiversity has also changed. The floristic richness has decreased. In 1978, 234 species were inventoried in the study area, for 134 in 2005 and only 94 in 2011. On the other side, the family composition and phytogeographical spectra have also changed, in the sense of an adaptation to xeric conditions. This can be explained by severe spells of drought combined by an exponential rise of livestock during the last 30 years. Finally, even if the last years were wetter and a slight improvement in vegetation cover was observed, there is no evidence for a desertification in reverse, as it is considered for the Sahel. On the contrary, we are still in a degradation process on the long term.

In: Arid and Semi-Arid Environments
Editor: Gabriel Ferguson
ISBN: 978-1-63483-383-7

Chapter 1

LINKING SEMIARID-MOUNTAIN AGRICULTURE TO ENVIRONMENTAL BENEFITS AND ADVERSE IMPACT

V. H. Durán Zuazo[1,*], C. R. Rodríguez Pleguezuelo[2], J. R. Francia Martínez[3], S. Cuadros Tavira[4] and I. F. García Tejero[1]

[1]IFAPA Centro "Las Torres-Tomejil," Alcalá del Río, Sevilla, Spain
[2]Earth and Life Institute/ Environmental Sciences, Université Catholique de Louvain. Louvain-la-Neuve, Belgium
[3]IFAPA Centro "Camino de Purchil." Granada, Spain
[4]Universidad de Córdoba, Campus de Rabanales Crta. Córdoba, Spain

ABSTRACT

The soil erosion, runoff and nutrient-loss patterns were monitored in a mountainside with olive (*Olea europaea* cv. Picual) plantations under conventional and conservation agricultural systems based in two different soil management: 1) conventional tillage and 2) non-tillage with barley (*Hordeum vulgare*) strips of 4 m width, respectively. The erosion plots, located in Lanjaron on the southern flank of the Sierra Nevada Mountains in SE Spain, had 30% slope at an altitude of 565 m and 192 m^2 (24 m x 8 m) in area. The highest erosion rate of 3.5 t ha^{-1} was registered under conventional system and runoff of 11.4 mm under conservation system,

* Corresponding author: victorh.duran@juntadeandalucia.es.

over the entire study period (19 erosive events). The conservation system reduced the soil erosion by 69% with respect to the conventional system, and the runoff under conventional system in relation to conservation system by 54%. The total NPK fluxes by runoff from conventional system amounted to 2.87, 0.09 and 0.93 kg ha^{-1}, and from conservation system 1.57, 0.09 and 0.98 kg ha^{-1}, respectively. In general, the soluble average concentrations of plant nutrients (N-NO_3, N-NH_4, P-H_2PO_4, and K) in the surface runoff were higher than the recommended level for standard water quality for from conventional system. These findings support the recommendation of conservation agricultural techniques for semiarid and sloped agricultural land with rainfed-tree crops for erosion and pollution control.

Keywords: Semiarid agriculture, runoff; erosion; olive orchards, mountainous land; non-point source pollution

1. INTRODUCTION

Since ancient times the olive tree (*Olea europaea* L.) is the most widespread crop of the Mediterranean area, growing under rainfed conditions on marginal lands, due to its ability to adapt to drought that is mainly cultivated in shallow, stony, and steep soils.

According to the FAOSTAT (2015), the olive tree covers a surface area of about 9.5 Mha in the world, of which more than 2 Mha of olives are grown in Spain. Concretely more than 75% of olive area is located in Andalusia (S Spain), where many orchards are confined to slopes or rugged land, and under different production systems (de Graaff et al. 2008).

The widely spaced olive trees on slopes combined with poor vegetation cover of the soil increases the vulnerability of the orchards to soil erosion (Panagos et al. 2015; Taguas and Goméz, 2015). Also, in this areas water is the most limiting resource, where the climate is typically characterized by high potential evaporation and low and highly variable rainfall during the growing season. Consequently, measures to control the soil erosion and runoff are crucial in order to mitigate these negative effects produced during the rainfall storms (Durán et al. 2011).

The traditional tillage can induce roughness and may have serious impact on runoff and erosion patterns, many olive orchards have used this technique in order to reduce the competition between trees and weeds for water uptake. There are various reports concerning the impact of soil management on soil erosion in olive orchards in Mediterranean zones (Hernandez et al., 2005; Gómez et al. 2009a, Palese et al. 2014), however, the results are contradictory due to the diverse set of environmental conditions considered.

Conservation agricultural techniques include: non-tillage combined with mechanical weed control; reduced tillage with or without weed control; and a cover crop sown in the autumn and killed with herbicides in the spring (Maetens et al. 2012). In such an environment, cultivating plants in strips running across the slope can lessen the erosion by reducing runoff and breaking up its trajectory over sloping lands, thus, unnecessary runoff losses are minimised by promoting its infiltration in the soil matrix (Durán et al. 2009).

The release of plant nutrients from soil and their transport in runoff, as is the case of agriculturally applied NPK, may affect the quality of the water bodies receiving these effluents, especially in areas that are located in lowlands (Ordoñez et al. 2007; Gómez et al. 2009b). Mountain agriculture for olive production involves fertilization, often applying more nutrients than required by crop; this excess could be contributed to environmental pollution, especially from runoff.

Nitrogen fertilizer is regarded as the main contributor to $N\text{-}NO_3$ losses by runoff that depends on drainage, the relative rate of plant uptake, and several N-transforming processes such as mineralization, immobilization, nitrification and denitrification. In contrast to the high capacity of $N\text{-}NO_3$ to transport, phosphorus $P\text{-}PO_4$ is rapidly retained as insoluble inorganic compounds and sorbed to soil particles, being the losses in surface runoff tend to be quite low. Potassium (K) is an essential plant nutrient in fruit production, and therefore olive growers, tend to apply heavy amounts of this element to encourage good-quality fruit and to boost the foliar content in K-deficient orchards.

The objective of this work was to compare the effects of two agricultural systems in steeply sloping areas on soil erosion, runoff, and plant-nutrient transport under the experimental conditions of Lanjaron (Granada, SE Spain). Concretely, conventional and conservation agricultural strategies in olive orchards based in two soil-management systems: 1) conventional tillage and 2) non-tillage with barley (*Hordeum vulgare*) strips of 4 m width, respectively.

Figure 1. Runoff-erosion plots used for the experiment in Lanjaron (SE Spain).

Table 1. Production systems used in the study

System	Soil management	Cultural operations			
		Winter	Spring[1]	Summer	Autumn[1]
Conservation	Non-tillage with barley strips	Weed control between strips with pre-emergence residual herbicide, diuron (1 kg ha^{-1})	Chemical kill of barley strips (mulch of barley straw)	None	Mouldboard ploughing to 5-cm depth. Sowing of barley in strips of 3 m wide (180 kg ha^{-1})
Conventional	Conventional tillage	None	Mouldboard ploughing to a depth 15 cm	None	Mouldboard ploughing to a depth 15 cm

[1] Fertilizer (N) by broadcasting in spring; [2] Fertilizer (NPK) by broadcasting in autumn. Fertilizer in the form of 15-15-15 NPK at a nominal rate of 453.2 kg ha^{-1} and urea 301.5 kg ha^{-1} yr^{-1}.

2. Materials and Methods

2.1. Site Description, Experimental Setting, and Field Measurements

The experiment was conducted in rainfed olive orchard of 30-year-old olive trees (*Olea europaea* cv. Picual) spaced 8 x 8 m in Lanjaron (Granada, SE Spain), with UTM coordinates X = 456,747.60; Y = 4,084,354.5 and at 565 m a.s.l.. The soils of the zone, Typical Xerorthent (Soil Survey Staff 1999), had loamy-soil texture with sand, silt, and clay of 540, 270, and 190 g kg^{-1}, respectively, 12.0 g kg^{-1} of organic matter, 1.1 g kg^{-1} of total N, 11.2 mg kg^{-1} available P and 51.8 mg kg^{-1} available K.

The closed erosion plots are located on a mountain slope of 30%, were 192 m^2 (24 m x 8 m) in area and contained 3 trees each. Figure 1 shows the erosion plot consisted of a galvanized enclosure, drawer collector (4 m x 0.55 m), and three sediment and runoff tanks (50 cm in diameter and 100 in height), being the first and second tank with a nine-slot divisor (1:9). Table 1 displays the treatments consisted in two soil-management systems were applied: conventional tillage and non-tillage with barley (*Hordeum vulgare*) strips of 4 m width (Figure 2).

Figure 2. Plant strips used in plots with conservation techniques in sloping olive orchards.

The rainfall data were collected from a local meteorological station (<100 m from the plots). For each storm, average intensity [I = (Total rain/total time) (mm h^{-1})] and maximum intensity at 30 min (I_{30}) were calculated.

During the study period the soil erosion and runoff from plots were collected and measured after each rainfall event. Three 1-liter aliquots were taken from different depths while the content of the tanks was being stirred. The sub-samples were immediately taken to the laboratory, oven dried at 105 °C and weighed to determine the suspended sediment concentration. In addition, each runoff sample was analysed for NO_3, NH_4, PO_4, and K concentration in accordance with standard methods for the examination of waters (APHA, AWWA, WPCF, 1995).

Linear relationships were fitted between runoff, soil loss, and rainfall parameters (rainfall depth and I_{30}).

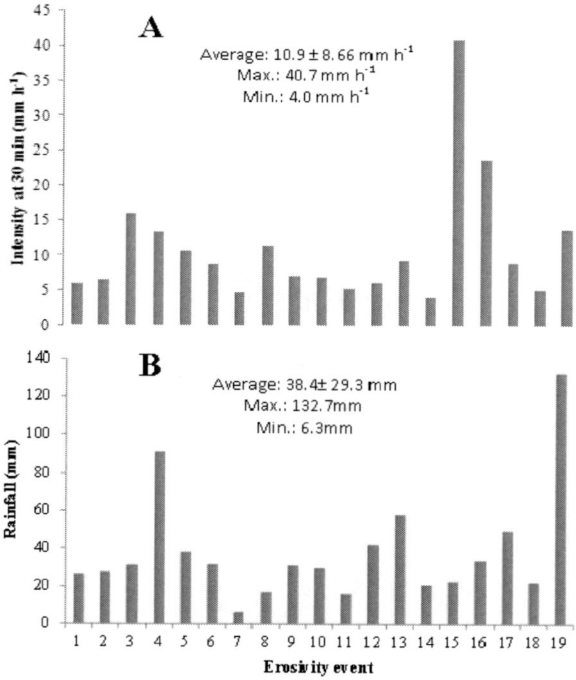

Figure 3. Maximum intensity at 30 min rainfall (A) and rainfall depth (B) during the 19 erosive events. ± Standard deviation.

Table 2. Average soil erosion and runoff for the different agricultural systems

	Conventional system		Conservation system	
	Soil erosion ($t\ ha^{-1}$)	Runoff (mm)	Soil erosion ($t\ ha^{-1}$)	Runoff (mm)
	Spring			
Total	3.03	1.0	0.53	1.9
Average	1.51 ± 2.14	0.5 ± 0.7	0.27 ± 0.36	1.0 ± 0.6
	Autumn			
Total	0.44	3.6	0.49	5.3
Average	0.15 ± 0.25	1.2 ± 1.8	0.16 ± 0.26	1.8 ± 2.1
	Winter			
Total	0.03	0.6	0.06	4.2
Average	0.01 ± 0.02	0.2 ± 0.3	0.02 ± 0.03	1.4 ± 1.6
Total amount	3.50	5.2	1.1	11.4

± Standard deviation

3. Results and Discussion

3.1. Rainfall and Soil Erosion and Runoff

Figure 3 displays the rainfall depth and its intensity (I_{30}) for the 19 events during the monitoring period, averaging 132.7 mm and 10.9 mm h^{-1}, respectively. In particular, the I_{30} during the study period ranged from 4.0 to 40.7 mm h^{-1}, showing high contrasts in both quantity and intensity (Figure 3A). In this sense, the frequency of I_{30} and rainfall depth for the monitoring period, these being the 84% between 4-15 mm h^{-1} and the 68% higher than 25 mm, respectively. This exhibits the typical seasonal pattern of the Mediterranean climate with rainfall concentrated in autumn and winter months (Figure 3B). This fact shows that the main vulnerability of the Mediterranean area to erosion are the high intensity of the rainfall after summer, and the high fluctuations over the short and long term, especially in rainfall quantity.

Table 2 shows the results for the effect of agricultural systems applied through the soil-management strategies on the average soil erosion and runoff. The highest average of soil loss was registered in the conventional system with tillage and without plant strips (3.50 t ha^{-1}), being those values lowered from the conservation system with non-tillage and barley strips (1.10 t ha^{-1}). For both systems the soil erosion was higher during the spring months due to heavy rainfall storms which are in line with the high values in I_{30} during this period. In relation to runoff, this was higher in conservation system amounted to 11.4 mm, being the highest runoff rates for both systems was fixed during the autumn months. Therefore, high variability of the rainfall rate and intensity caused a wide range in the amount of soil loss and runoff, with the highest values in May, October, and December, in line with the highest rainfall depth and intensity for the monitoring period.

The findings show that the conservation system markedly reduced soil erosion about 69% in relation to conventional. The rainfall energy reaching the soil surface was dissipated by barley strips, the rainfall water being diverted to stemflow, buffering the mechanical impact of the raindrops, diminishing the splash and soil transport. That is, the plant covers in intercepting rainfall is a determining factor for controlling soil erosion, avoiding the development of impermeable crusts at the soil surface. In addition, the barley roots, by penetrating the soil matrix favoured porosity and thus infiltration, allowing the root zone to act as a partial sink for runoff. However, during the rainy period when soil-water content is high, the excess of runoff can run perpendicular to the slope after trapping eroded soil by plant strips.

Under the conventional system the autumn tillage and part of winter could reduce the amount of soil erosion and surface runoff by temporally increasing the roughness of the soil surface, that is, the runoff from tilled plots declined sharply after tillage but rose again following a number rainstorms without tillage. Consequently, the benefits of the conventional system are effective in a short term.

The results of the present study showed that conservation system with barley strips across the slope could reduce erosion than conventional system. In addition, the mulch cover after spraying or cutting the plants improve soil physico-chemical parameters, facilitate soil-water and soil biological activity, aid in weed control, and help recycle plant nutrients. Consequently, this control is particularly urgent in traditional olive-producing areas, where orchards cover vast tracts of land.

Our results agree with the experiments of Raglione et al. (1999) measured a greater erosion rate under conventional system than with a permanent cover (82.8 and 0.7 t ha^{-1}, respectively). In line with this, Kosmas et al. (1996) reported a noteworthy reduction of soil erosion (0.03 t ha^{-1}) under natural spontaneous vegetation cover in olive orchards.

The relationships among rainfall characteristics, soil erosion and runoff were estimated. Concretely, the soil erosion was highly dependent on the rainfall intensity I_{30} for both conventional (R^2 = 0.89) as well as for conservation (R^2 = 0.86) systems. However, the runoff was correlated with rainfall depth, especially for conservation system (R^2 = 0.79). By contrast, soil erosion and the runoff were not correlated with rainfall depth and rainfall intensity I_{30}, respectively. Despite the great variability in rainfall and intensity events for the monitoring period, soil erosion was dependent on runoff, this trend being more pronounced for the conservation (R^2 = 0.57) than for conventional system.

In short, the fit of the rainfall parameters characteristics versus soil erosion and runoff showed a significant correlation in the study zone, especially for the conservation agricultural system.

3.2. Plant-Nutrient Transport By Runoff

Table 4 presents the results for the nitrate, ammonium, phosphate and potassium transport per area by runoff for the two agricultural systems (discharge-weighted).

Table 3. Nutrient fluxes with runoff (discharge weighted) under different agricultural systems

	$N\text{-}NO_3$	$N\text{-}NH_4$	$P\text{-}PO_4$	K
	(mg m^{-2})			
Conventional system				
Average	100.3 ± 211.5	7.8 ± 5.9	2.8 ± 2.6	9.3 ± 6.5
Maximum	658.7	17.3	8.1	19.1
Minimum	0.5	1.2	0.4	1.9
Total	1,003.2	77.8	28.3	92.9
Conservation system				
Average	22.5 ± 25.5	5.8 ± 7.3	1.7 ± 2.5	5.7 ± 7.1
Maximum	74.8	24.8	7.8	22.2
Minimum	0.3	0.0	0.0	0.1
Total	359.3	98.0	27.7	97.6

± standard deviation.

The greatest total $N\text{-}NO_3$ transport was from the conventional plot amounted to 1,003.2 mg m^{-2}, with a decreasing trend from conservation system with 359.3 mg m^{-2}. Both $N\text{-}NH_4$ and K followed a similar pattern (98.0

and 97.6 mg m^{-2}, respectively), the greatest flux being from the conservation plot. In relation to P-PO_4 was not found almost any difference.

The highest N-NO_3 and N-NH_4 concentrations in runoff was fixed under conventional system amounted to 25.5 and 7.3 mg l^{-1}, respectively. The variability in concentration of different forms of N is not surprising, due to N-cycling processes that can cause rapid transformations from one form to another, and temporal as well as spatial variability of N applications.

The higher N-NO_3 and N-NH_4 content from the runoff in conventional plots indicated a potential source of inorganic N pollution. However, the seasonal average concentration of N-NO_3 from both systems was well within the 50 mg l^{-1} upper limit recommended for water (WHO 2008). In contrast to the N-NH_4 concentrations, especially the conventional treatment exceeded the 0.5 mg l^{-1} standard for water supplies (WHO 2008). In addition, there were several runoff events from all plots where N-NH_4 concentration even exceeded the 2 mg l^{-1}, a level considered toxic to fish.

The maximum P-PO_4 concentrations for both conventional and conservation systems was of 2.70 and 0.90 mg l^{-1}, respectively, exceeding established limits associated with eutrophication of surface waters: from 0.01 mg P l^{-1} (Vollenweider and Kerekes, 1980) to 0.05 mg l^{-1} (U.S. EPA, 1976), or 0.05-0.1 mg l^{-1} total P for the protection of fresh water (ANZECC, 1992). Although the phosphate is not toxic, extremely high concentrations present a certain risk for the productivity of aquatic ecosystems. For the protection of such systems, MacDonald et al. (1991) proposed limits of 0.025 to 0.1 mg l^{-1}, a range exceeded by the P concentration in our experiment.

The K concentrations in plots with both systems were within the 12 mg l^{-1} upper limit recommended for drinking water (Griffioen 2001).

The dominance of N-NO_3 suggests that the dissolved nitrogen in the runoff came mainly from N fertilizers applied in the field, rather than from the soil, and the same applies to the dissolved NH_4^+ contained in the runoff, especially from rapid urea hydrolysis. Similarly for the K-rich runoff, this presumably was from K fertilizers used for fruit production.

The leaching and transport of plant nutrients by runoff from rains caused progressive loss of the soil fertility as well as the contamination risk of the both surface water and groundwater (Rodriguez et al. 2014). The agricultural conservation techniques proved more effective compared to conventional in retaining soil nutrients from transport by runoff.

We concluded that the implementation of conservation techniques in olive orchards on a 30% slope in the framework of this experiment substantially controlled soil erosion and runoff. In this context, the plant strips and their cover mulch between rows of olive trees proved to be an effective system for trapping runoff and eroded soil in comparison to the conventional system. In addition, the plant strips shielding the soil surface from raindrop impact while concentrating greater amounts of plant litter, which improves the soil matrix through mineralization and humification processes.

On the other hand, the conservation system reduced the plant-nutrient losses (NPK) in comparison to the conventional, even though soluble K concentration was slightly greater than with conventional techniques. This system not only produced greater nutrient losses from olive orchards but also represented an important source pollution of the study area. Thus, the results of this study highlight the potential conservation benefits of plant strips to control soil erosion, runoff and plant-nutrient losses.

ACKNOWLEDGMENT

This publication was sponsored by the following research project "Agro-environmental assessment of conventional, integrated, conservation and organic olive production systems in mountainous areas: impact on erosion, soil quality and commercial value of olive oil" (P11-AGR-7431) granted by CEICE of Junta de Andalucia.

REFERENCES

ANZECC. (1992). Australian Water Quality Guidelines for Fresh and Marine Waters. Australian and New Zealand Environment and Conservation Council.

APHA, AWWA, WPCF. (1995). Standard Methods for the Examination of Water and Wastewater. 17th ed, Washington DC., USA.

de Graaff, J., Durán, Z. V. H., Jones, N. & Fleskens, L. (2008). Olive production systems on sloping land: Prospects and scenarios. *Journal of Environmental Management*, *89*, 129-139.

Durán, Z. V. H., Rodríguez, P. C. R., Arroyo, P. L., Martinez, R. A., Francia, M. J. R. & Cárceles, R. B. (2009). Soil Conservation Measures in Rainfed Olive Orchards in South-Eastern Spain: Impacts of Plant Strips on Soil Water Dynamics. *Pedosphere*, *19*., 453-464.

Durán, Z. V. H., Rodríguez, P. C. R., Flanagan, D. C., García, T. I. & Muriel, F. J. L. (2011). Sustainable land use and agricultural soil. In *Alternative Farming Systems., Biotechnology., Drought Stress and Ecological Fertilisation.*, Sustainable Agriculture Reviews 6, (ed. Lichtfouse., E.), 107-192., Springer Science + Business Media BV, The Netherlands.

FAOSTAT. (2015). Website: http://faostat.fao.org (website visited on May, 2015)..

Gómez, J. A., Sobrinho, T. A., Giráldez, J. V. & Fereres, E. (2009a). Soil management effects on runoff., erosion and soil properties in an olive grove of Southern Spain. *Soil and Tillage Research*, *102*, 5-13.

Gómez, J. A., Gema, G. M., Giráldez, J. V. & Fereres, E. (2009b). The influence of cover crops and tillage on water and sediment yield., and on nutrient., and organic matter losses in olive orchard on a sandy loam soil. *Soil and Tillage Research*, *106*, 137-144.

Griffioen, J. (2001). Potassium adsorption ratios as an indicator forthe fate of agricultural potassium in groundwater. *Journal of Hydrology*, *254.*, 244–254.

Hernández, A. J., Lacasta, C. & Pastor, J. (2005). Effects of different management practices on soil conservation and soil water in a rainfed olive orchard. *Agricultural Water Management*, *77*, 232-248.

Kosmas, C., Moustakas, N., Danalatos, N. G. & Yassoglou, A. T . (1996). The Spata field site. 1. The impacts of land use and management on soil properties and erosion. 2. The effects of reduced moisture on soil properties and wheat production. In: Brandt JC, Thornes J, editors. Mediterranean desertification and land use. Wiley & Sons & New York., USA, 207-228.

Maetens, W., Poesen, J. & Vanmaercke, M. (2012). How effective are soil conservation techniques in reducing plot runoff and soil loss in Europe and the Mediterranean?. *Earth-Science Reviews*, *115*, 21-36.

MacDonald, L. H., Smart, A. W. & Wissmar, R. C. (1991). Monitoring guidelines to evaluate effects of forestry activities on streams in the Pacific Northwest and Alaska, US Environmental Protection Agency Report, LPA 910/9-91-001.

Ordóñez, F. R., Rodríguez, L. A., Espejo, P. A. J., González, F. P. & Saavedra, M. M . (2007). Soil and available phosphorus losses in ecological olive groves. *European Journal of Agronomy*, *27*, 144-153.

Panagos, P., Borrelli, P., Meusburger, K., Alewell, C., Lugato, E. & Montanarella, L. (2015). Estimating the soil erosion cover-management factor at the European scale. *Land Use Policy*, *48*, 38-50.

Palese, A. M., Vignozzi, N., Celano, G., Agnelli, A. E., Pagliai, M. & Xiloyannis, C. (2014). Influence of soil management on soil physical characteristics and water storage in a mature rainfed olive orchard. *Soil and Tillage Research*, *144*, 96-109.

Raglione, M., Toscazo, P., Angelini, R., Briccoli, BC., Spadoni, M., De Simone, C. & Lorenzini, P. (1999).. Olive yield and soil loss in hilly environment of Calabria (Southern Italy). Influence of permanent cover crop and ploughing. In: Bech J, editor. Analysis of from a physical based model. 6-th *International Meeting on Soils with Mediterranean Type of Climate. Barcelona*, Spain, 1038-1040.

Rodriguez, G. V., Mendes, M. P., Garcia, S. M. J., Chica, O. M. & Ribeiro, L. (2014). Predictive modeling of groundwater nitrate pollution using Random Forest and multisource variables related to intrinsic and specific vulnerability: A case study in an agricultural setting. (Southern Spain). *Science of The Total Environment*, 476–477, 189-206.

Soil Survey Staff. Soil Taxonomy. (1999). A Basic System of Soil Classification for Making and Interpreting Soil Surveys., Agric. Handbook N 436., USDA-USA.

Taguas, E. V. & Gómez, J. A. (2015). Vulnerability of olive orchards under the current CAP (Common Agricultural Policy). regulations on soil erosion: a study case in Southern Spain. *Land Use Policy*, *42*, 683-694.

Vollenweider, R. A. & Kerekes, J. (1980). The loading concept as a basis for controlling eutrophication philosophy and preliminary results of the OECD programme on eutrophication. *Prog Water Technol*, *12*, 5-38.

WHO. (2008). Guidelines for drinking-water quality. World Health Organization. 3rd ed. Incorporating the first and second Addenda. Volume 1, Recommendations. Geneva, Switzerland.

In: Arid and Semi-Arid Environments
Editor: Gabriel Ferguson

ISBN: 978-1-63483-383-7

Chapter 2

THERMAL IMAGING TO ASSESS THE PLANT-WATER STATUS IN WOODY CROPS UNDER ARID AND SEMI-ARID CONDITIONS

Iván Francisco García-Tejero**, *Víctor Hugo Durán-Zuazo* and *José Luis Muriel-Fernández
Centro IFAPA "Las Torres – Tomejil. Ctra. Alcalá del Río, Sevilla, Spain

ABSTRACT

Water is the most limiting factor for irrigated agriculture in arid and semi-arid areas of European Mediterranean countries. The searching of alternative irrigation strategies such as deficit irrigation and the use of different tools to monitor the crop-water status are some options to improve the water management under limiting drought conditions. Infrared thermography is a non-invasive promising technique to monitor the crop-water status based on the relations between the leaf or canopy temperature and the transpiration rate, which is largely related with the stomatal conductance. From an energy balance point of view, as a consequence of stomatal regulation (partial closure) under a mild or moderate water stress, leaf temperature tends to increase because of a decrease in the heat dissipation associated with the transpiration process,

* E-mail corresponding author: ivanf.garcia@juntadeandalucia.es

which is known as the evaporative cooling process. However, temperature readings are highly dependent on many meteorological variables being very difficult to define threshold values of canopy temperature for the irrigation scheduling or the crop-water status monitoring. For this reason, in spite of many works devoted to the use of infrared thermography, there are many questions and doubts that have to be solved for real optimization, protocolling and standardization of this technique, (such as, when the images should be taken, the best time along the day, the most representative, and relevant water stress index or the image processing). Taking into account these considerations, the aim of this work was to respond to some questions that could help to optimize the use of this technique in some woody crops: i) which are the main advantages and disadvantages of this technique?; ii) which of the actual thermal indices is the most useful and representative?; iii) when is the best moment to assess the crop physiological status using the canopy temperature, focusing the answers in woody crops.

Keywords: Thermography, thermal indices, crop-water monitoring, deficit-irrigation scheduling

1. Introduction

Water is the major limiting natural resource for crop production in irrigated agriculture, especially in arid and semi-arid areas of European Mediterranean countries such as Spain [1]. Many studies have been devoted to increase the actual knowledge for irrigation scheduling under water scarcity situations, most of them based on plant responses monitoring to drought. In these sense, plants are classified into two major groups: iso- and anisohydrics, depending on the capacity to regulate their transpiration as function of soil water availability [2].

Stomatal regulations of gas exchange in response to drought have a great importance in the plant adaptation to non-optimal situations [3]. This regulation induces to keep equilibrium between photosynthetic rate, water loss, and leaf temperature [4]. On the one hand, a moderate water stress can promote a water saving together with an improvement in some quality parameters of yield in many crops [5, 6] On the other hand, a severe water stress induces a stomatal closure, promoting a significant reduction of assimilative activity and hence, significant yield reductions and pernicious effects in crop development can be promoted [7]. In addition stomatal regulations in response to drought have a great importance in the plant adaptation to non-optimal situations. This regulation induces to keep an equilibrium between the photosynthetic rate, water loss and leaf temperature [8]. Many woody crops such as olives (*Olea europea* L.), citrus (*Citrus sinensis* L.), almond (*Prunnus dulcis* Mill.) or grapevines (*Vitis vinifera* L.) among others are able to keep the crop-water status within certain threshold values when these are subjected to mild water stress situations, promoting water savings, even together with improvements in some quality parameters of yield [4]. Because of this, there is a strong interest in developing and improving different robust tools able to monitor the plant-water status, with the aim of keeping the crop within of certain threshold values of moderate-to-mild water stress.

2. Remote Monitoring of Crop Water Status

After the green revolution developed at 50s of last century, today is occurring a second great technological revolution in the agriculture, which is bringing a progressive modernization of agricultural activities [9].

Figure 1. False coloured images of different plants and at different levels: a) *Arabidopsis thaliana*; b) vineyards, c) stem of almond; d) strawberry; e) citrus f) almond trees.

The introduction of new technologies on agricultural systems has not only improved the productivity of agricultural systems, but also provide answers to situations for which a few years ago there were no alternatives. An example is the use of remote sensing systems for monitoring crop water status [10]. These systems rely on the detection and recording of spectral variations of crops both visible radiation range, and within the electromagnetic spectrum in the infrared range [11]. Remote sensing is being increasingly used for the development of studies on plant physiology and practical applications in agriculture and forestry, because of the provided information is related to several morpho-physiological parameters (biomass, transpiration, and photosynthesis) that characterize the level of crop development and crop physiological status [12]. The advantages of remote sensing are many: it is a non-destructive, rapid, non-invasive method, and allows work at different scales (from seedlings to woody crops and large crop areas (Figure 1), which allows data logging and real-time, obtaining quantitative information on the variability of crops.

2.1. A Brief Introduction to Thermal Imaging

The infrared thermography is a technique used in many disciplines such as medicine, industrial applications (automotive, aerospace, cold machines) construction, and agriculture, among others. Its base is the study of the emitted infrared radiation emitted by the bodies, and all those properties related to this physical process [13]. The infrared radiation corresponds to the radiation between the spectrums of light visible and the microwaves, comprising wavelengths between 0.75 and 1000 μm (Figure 2).

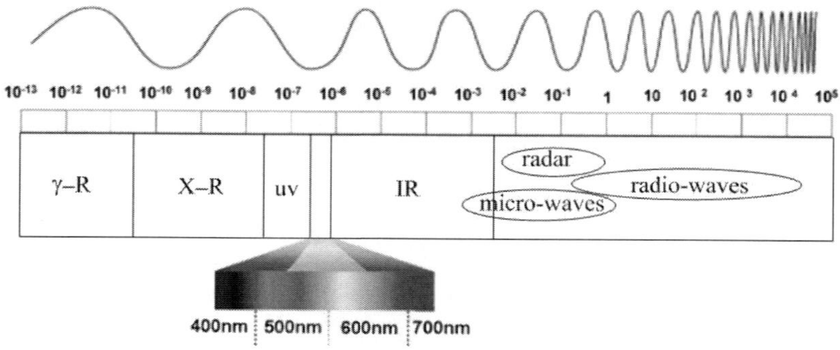

Figure 2. Scheme of electro-magnetic spectrum.

Any body has the ability to capture and exchange energy with the surrounding environment, through direct contact; convection (through a medium; liquid or gas); and radiation (through the electromagnetic spectrum). When we think about "heat," we are referring to thermal energy in transition, while the temperature is a property of matter, which is directly related to the ability of a body to emit a certain radiation like heat; as it is defined by the Stefann-Boltzmann law:

$$W = \varepsilon\sigma T^4 \qquad \text{(Eq. 1)}$$

where W is the total radiation emitted by a body per unit area (w m^{-2}); ε, emissivity; σ, the Stefann-Boltzmann constant (5.67×10^{-8} W $m^{-2}K^{-4}$); and T, the surface temperature of a body.

The Stefan-Boltzmann law is applied to black bodies; that is, perfect bodies in terms of radiation emitters; which have an emissivity (ε) equal to 1. The emissivity is defined as the ability to radiate/transmit energy of a body. Real bodies have $\varepsilon<1$, this ranging in the case of plants, between 0.92 and 0.98.

The total infrared radiation detected by an infrared sensor is the sum of the emitted and reflected radiation by the observed object (Equation 2), considering as not-significant the absorption of infrared radiation by the mass of air between the studied object and the camera sensor, when the distance is not large (a few meters) or when the infrared sensor works in the range of 3-5 or 7-14 μm, where the transmissivity of the atmosphere to infrared radiation is high [14].

$$W = \varepsilon\sigma\,(T_s)^4 + (1-\varepsilon)\,W_{background} \qquad \text{(Eq. 2)}$$

When the distance between the infrared sensor and the studied object is greater, it is necessary to introduce another factor; the atmospheric emission (W_{ATM}), and the transmissivity of the atmosphere to radiation (τ) according to equation 3:

$$W = \tau\,[\varepsilon\sigma\,(T_s)^4 + (1-\varepsilon)\,W_{background}] + W_{atm} \qquad \text{(Eq. 3)}$$

Nowadays, most of thermal imaging cameras already include this type of correction, allowing the development of more accurate measurements of the temperature of the observed objects. However, it is necessary to introduce the values of emissivity, relative humidity and air temperature, the distance to the object and the reflected temperature by the surrounding environment. This last parameter can be estimated by measuring the temperature reflected by an irregular piece of aluminium foil, placed between the camera and the object to be measured and used to this an emissivity value equal to 1.

2.2. How Does a Thermal Imaging Camera Work?

During the process of reception and transmission energy, part of the incident radiation is transmitted by the body (τG) and some is absorbed by the body (αG). Furthermore, part of the incident radiation is directly reflected by the body (ρG); while the radiation that has been absorbed previously, a fraction of this is issued (E). The sum of these latter two items (ρG + E) is known as radiosity (J), which is the total electromagnetic radiation emitted by the bodies (Figure 3), this being detected by the infrared sensor [8].

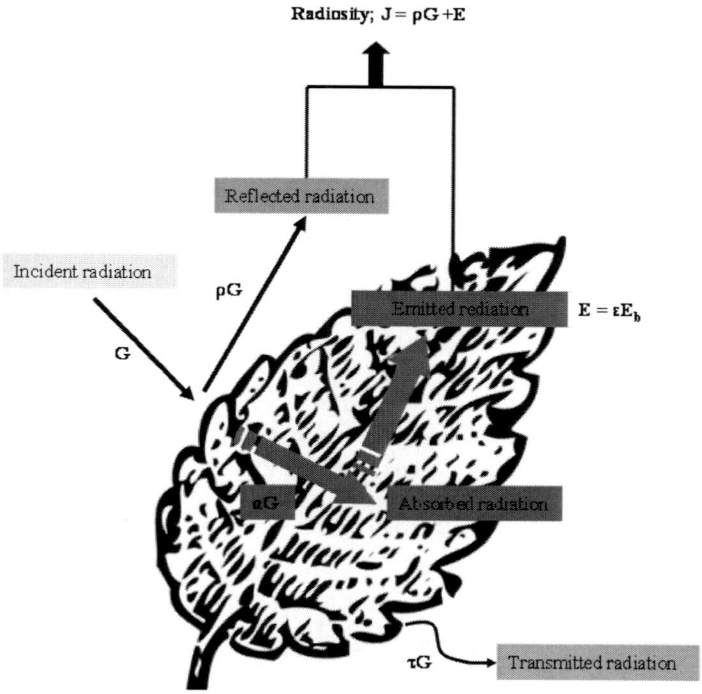

Figure 3. Scheme of balance energy in a leaf.

However, when we work with these kinds of tools, we must consider some parameters that will influence directly during the time measuring. Not all the received radiation by the camera sensor corresponds to the radiosity of the body or the element with which we are working; but there is other components such as a fraction of radiation from the medium between the measuring object and the electromagnetic sensor, and other corresponding to the energy reflected by the result of radiation from the surrounding environment (Figure 4). These issues that at the first could be complicated, today are largely outweighed by the software and the algorithms transformation from which the machinery is able to adjust and re-calculate the object temperature as only with a few known additional variables such as the distance to the object to be measured, the value of the reflected temperature, relative humidity, the angle of inclination or the emissivity of the body with which we are working [8].

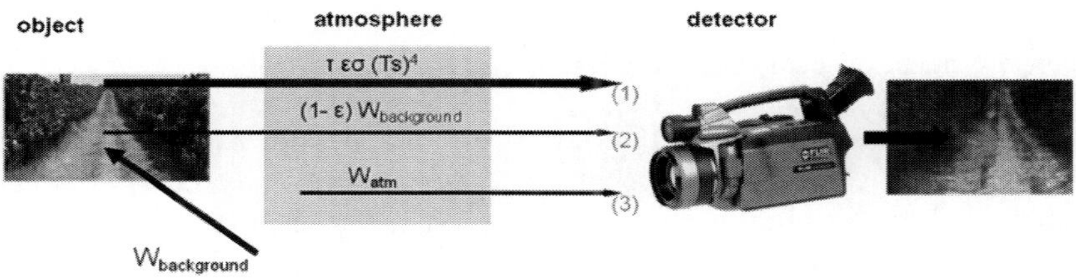

Figure 4. Emission-reception process of the radiation detected by the sensor.

2.3. Plant-Environment Interactions

Crops interact with the environment in different ways, mainly through a matter exchange (CO_2, H_2O, mineral salts, etc.); and through an energy exchange (light in the process of photosynthesis, and as heat). These interactions may vary due to mutations or changes in a situation of biotic or abiotic stress. As for the capacity of a crop to exchange energy with the environment, it is capable of absorbing energy in form of short wavelength (light energy metabolism involved in photosynthesis) and long-wave form (infrared). In turn, associated with transpiration process, exists a phenomenon known as "*evaporative cooling,*" in which there is a heat dissipation, associated with the loss of water vapour from the stoma [15]. Finally, there is a loss of energy as long wave through convection and conduction mechanisms. Of these processes, evaporative cooling is the most related with the crop-physiological status in terms of stress and crop temperature (Figure 5).

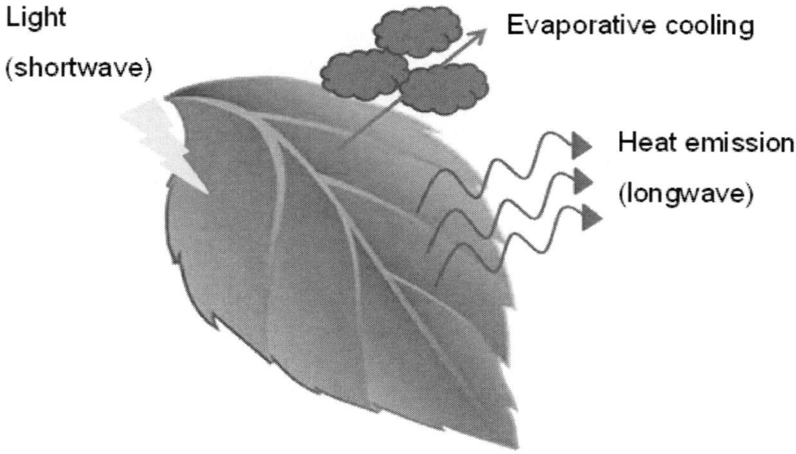

Figure 5. Scheme of energy balance at leaf level.

When a crop is being well irrigated, or if it has the amount of water required for its correct development, there is a vertical movement of water from the soil to the aerial part, going out as vapour through stomata. Under a situation of water scarcity, many crops have the ability of producing chemical signals (synthesis of abscisic acid ABA) that travels from the roots to the leaves triggering a response that ends with a total or partial closure of stomata. This fact causes a decrease in water loss from the plant tissues, thus avoiding compromising situations related to dehydration of the tissues or the total or partial embolism in the xylem vessels [9].

Because of this, there is a decline in CO_2 fixation, and therefore, the photosynthetic activity of the plant, but certainly more pernicious situations that could result in death of the crop are avoided [16]. Associated with the decrease in transpiration levels, the process of evaporative cooling is significant reduced, increasing the levels of leaf temperature typically several degrees above the air temperature [15].

These relationships establish the base for the crop-water status monitoring based on leaf/canopy temperature, since any situation that is causing a partial closing of the stomata (for example by a water-stress situation more or less prolonged) will result in a significant increase in leaf temperature thereof.

3. Why Is Necessary the Use of Thermal Indexes to Monitor the Crop Water Status?

Considering all the exposed above, thermal imaging emerges as a robust, reliable and non-invasive technique for assessing the crop-water status and irrigation scheduling [17, 18]. From an energy balance point of view, as it has been previously stated, as a consequence of stomatal regulation (partial closure) under a mild or moderate water stress, leaf temperature trends to increase, because of a descend in the heat dissipation associated with the transpiration process [11, 12, 19], which is known as the *evaporative cooling process*. This is a heat loss associated to transpiration process from the canopy so that a transpiration decline is traduced into an increase of leaf temperature [20].

At this point, it is necessary to launch the following question: Is the leaf temperature exclusively dependent on the crop transpiration level, or are there other variables that can significantly influence the absolute value of this term? According to Jones et al. [19] and Jones and Vaughan [11], there are many variables that will determine the final value of the leaf temperature, and therefore, must be taken into account when making an assessment of crop-water status using the values of temperature/canopy temperature. In this sense, variables such as the radiation level, air temperature, vapour pressure deficit, the relative humidity or the angle of the radiation incident on the leaf surface will influence decisively on the absolute value of the leaf/canopy temperature.

Therefore, to control these limitations various stress indexes have been developed intended to minimize the effect of these variables in studies related to the temperature of the crops and their relationship with the response to water stress. These indexes try to normalize the absolute values of temperature, obtaining a second value in which the effects of this set of potentially influential variables are partially minimized (Table 1).

Also, both the rate of crop water stress (CWSI) as the I_G require the use of reference values; that is, T_{wet} (referring to the leaf temperature value when plant is full irrigated and the transpiration level is maximum) or T_{dry} (leaf temperature value when stoma are totally closed); are used to obtain some stress indices which allows to normalize the obtained values of leaf temperature [8, 19, 21]. Within these indices, the CWSI, the thermal index to relative stomatal conductance (I_G) or ΔT, (this being the difference between canopy and air temperature) are between the most commonly used. In the case of T_{dry} these values are typically obtained covering one or two leaves with vaseline on both sides (upper and back sides) at least 20 minutes before taking pictures; while for the case of T_{wet}, this value is obtained spraying with water one or more leaves at the instant of measurement, or using wet porous materials (Figure 6).

Table 1. Most used thermal indexes in the crop water status monitoring

Index	Equation	Reference
CWSI	$CWSI = \frac{(T_{canopy} - T_{wet})}{(T_{dry} - T_{wet})}$	[3, 12, 22]
I_G	$I_G = \frac{(T_{dry} - T_{canopy})}{(T_{canopy} - T_{wet})}$	[12, 21, 23]
ΔT	$\Delta T = T_{leaf} - T_{air}$	[22, 24]

T_{dry}, Leaf temperature when a crop is under a maximum water stress situation; T_{wet}, leaf temperature when a crop is not under water stress (full irrigated); T_{leaf}, leaf temperature; T_{air}, air temperature at the moment of data acquisition.

Figure 6. Taking measurements position (a) and false-coloured image taken with the ThermaCam, in which two leaves have been previously selected to establish the T_{dry} and T_{wet} values.

However, in spite of many works have been devoted about the use of infrared thermography to irrigation scheduling and crop-water status monitoring, up to now, there are many questions and doubts that have to be solved for a real optimization, protocolling and standardization of this technique, such as, when the images should be taken, the best time along the day, the most representative and relevant water stress index, or the aspects related to image processing.

4. THERMAL IMAGING APPLICATION TO WOODY CROPS

Although any crop has a set of inherent characteristics; when we want to apply the thermal imaging to monitor the crop-water status, some common questions have to be taken into account. The first of them has to do with the *crop morphology*. There are some limitations usually associated with the procedure of capturing and processing data, requiring in many cases the use of complex software, thus reducing the operational and affordability of such these techniques. Taking an image is not a simple action; that is, you have to consider that not all the parts of the plant are susceptible of being monitored. The better the process of image capturing, more easy and precise will be the subsequent analysis of them.

Hand-operated cameras allow taking images of individual plants or portions of them, but with the disadvantage that, during the capturing process different elements (soil, shady areas, sky or portions of adjacent plants) can be reflected, requiring subsequent time-consuming processing images. This difficulty is specially marked in woody crops, with discontinuous canopies and a ground cover less than 100% [25]. Several authors such as Zarco-Tejada et al. [26], Wang and Gartung [27], or Jones et al. [12] have described different methods for overcome such limitations, although all of them have previously required different images processing, either through editing software and image processing, either through processes of classification of pixels, or through relatively laborious statistical analysis.

Figure 7 represents different examples of images in some woody crops. Within an image, there are many pixels that correspond to areas that must be avoided; such as branches, soil, sky, or other materials. Even the background will be a problem. That is, if we are monitoring a single plant, we are not interested in measuring the adjacent plants. However, many times it is not possible to avoid some parts of the plant during the monitoring process.

García-Tejero et al. [28] developed a single software able to eliminate those pixels corresponding with those areas that did not correspond to sunny leaves. According to these authors, images were taken from the sunny side of the canopy, avoiding the background problem using a cooled white screen, in order to simplify the isolation of the canopy surface through image processing.

Thermal images were processed with digital image processing tools. The raw thermal images were obtained in FLIR Systems' proprietary format by means of ThermaCamExplorer software (FLIR QuickReport) as a 'TIFF' image. The images were then processed, using the FLIR software for exporting these images since a TIFF to a 'CSV' file (comma separated values). Thus, having saved the CSV file corresponding to an image, it was made simple filtering, eliminating for each 'CSV' file the temperature values that did not correspond with a certain threshold and thus, leaving the pixels from the plant with its temperature values.

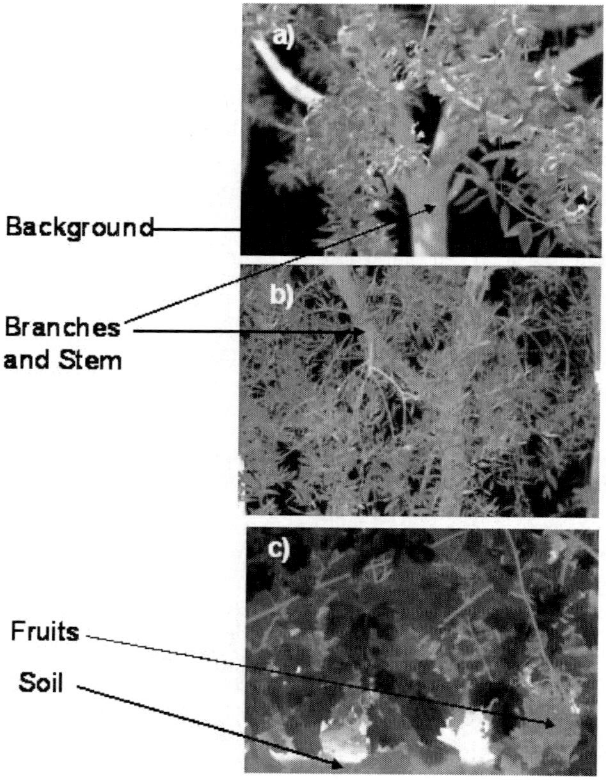

Figure 7. Views of different crops a) almond; b) olive; c) vineyard with different areas that must be avoided during the taking images.

The improvement obtained using these kinds of tools to eliminate those pixels/areas that do not correspond with leaves was very appreciable. In this line, making a comparison between two relationships between the values of ΔT vs. g_s in olives; the first of them obtained within a previous analysis of the images (Figure 8a); and the remaining, applying the single software developed by García-Tejero et al. [12], and selecting only those areas corresponding to the sunny leaves (Figure 8b), there was obtained a significant improvement in these results.

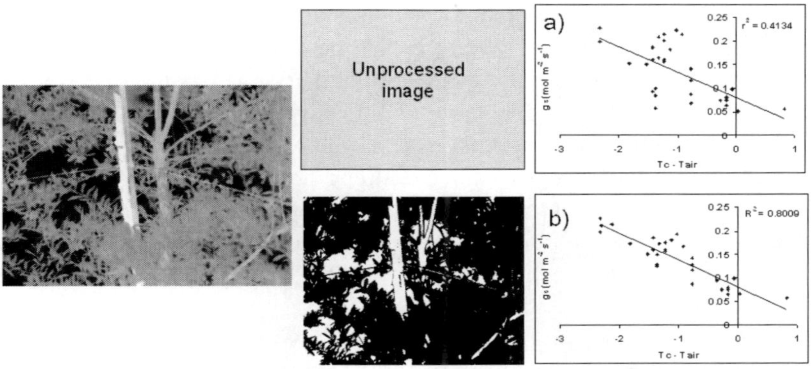

Figure 8. Differences in the relationships obtained without processing (a) and with image processing (b).

Other important question is related to the measuring time. Different experiences have been developed in this sense. Comparing different relationships between the ΔT index and some physiological measurements (leaf water potential; stomatal conductance, and net photosynthesis) obtained in measurements taken in the morning, at midday-early afternoon; and in the evening, it can be concluded that the best moment to monitor the crop-water status using the values of leaf or canopy temperature, would be during the hours of maximum solar radiation; between 12:00 and 15:00 h, independently of the monitored crop, as it is shown in Table 2 for citrus, almond and vineyard, respectively, has it has been previously discussed by other authors [17]. Many authors have reported similar answers to this question. Pou et al. [17] obtained the most significant relationships between the thermal indices CWSI and I_g with the g_s in vineyards, when these measurements were done at midday, suggesting that the reported information by the thermal images were more complete and sensitive at this moment of the day (r^2 = 0.75 for the relationships between g_s vs. I_g at 13:00 h; being this value of 0.49 and 0.61 for the taken measurements at 10:00 and 16:00 h, respectively); these data being in agreement with other reported by many authors such as Jones et al. [29]; Möller et al., [5]; Grant et al., [23]; Zia et al. [18] or Fuentes et al. [15], concluding that the thermal images should be taken at the early afternoon.

Finally, the third aspect would be related to the question about the best thermal index it is not so clear the relative question to the best thermal index in order to define the main relationships with the measured physiological parameters. In this sense, it is not so clear what of them could be the most robust, or what is offering more information about more information about the crop-water status. When we have to decide between them it is necessary to consider the advantages and disadvantages in their applications. According to the equations in Table 1, the most 'use-friendly' would be the ΔT index, because of we only need to know the absolute value of air temperature. However, this index changes with the atmospheric conditions, such as the VPD. Because of this, there are many studies available to modify the Tc to exclude the effects of environment such as using wet and dry reference leaves (as it was stated in the thermal indexes of CWSI or I_g, or using non-water-stressed baselines.

Table 2. Pearson's correlation coefficients obtained between ΔT index and g_s using measurements obtained at three moments along the day in three different woody crops

	Morning	Midday / Early afternoon	Evening
Citrus	*r* = 0.47*	*r* = 0.85**	*r* = 0.59*
Almonds	*r* = 0.46*	*r* = 0.84**	*r* = 0.42
Vineyards	*r* = 0.63*	*r* = 0.88**	*r* = 0.62*

* and ** show signification levels $p<0.05$ and $p<0.01$, respectively.

For the cases of CWSI and I_g [30, 31, 32], it is necessary not only the absolute value of leaf or canopy temperature, but, additionally, other reference values corresponding to the temperature of a well-irrigated and a non-irrigated crop. These values can be assessed, well directly taken from the field (in this case being necessary to have some plants under these conditions), or using baselines for a well irrigated crop. Obtaining these relationships is very complex, mainly in arid and semi-arid regions, because there is a high sensitivity to several climate factors such as radiation and wind speed [29, 33], which are not considered in the CWSI determination as it has been reported by Möller et al. [5]. Therefore, it is necessary to establish crop-specific nonwater-stressed baselines for different areas. This problem could be solved using a reference values of T_{wet} and T_{dry} throughout reference surfaces, for instance, using a wetted or fully transpiring section of the canopy, as it was proposed by many authors such as Jones et al. [34]; Leinonen an Jones [35] or Costa et al. [8], among others.

In spite of this, this option is not clearly defined, as it was stated by Meron et al. [36], focusing the difficulties to obtain an uniform protocol, and the difficulties linked to reproducibility. This fact could be solved, using wet artificial reference surfaces for T_{wet} and estimating T_{dry} as the sum of T_{air} + 5°C as it had been previously proposed by Irmak et al. [37].

Consequently, the selection of one these index should be oriented not only from the point of view of data acquisition and time-consuming, but the robustness of them. It is not easy to define this fact, because as it has been previously described, numerous interferences and variables could affect to the canopy temperature values.

From the energy balance point of view, leaf temperature is inversely correlated with transpiration (Figs. 3 and 5), and hence, with the stomatal conductance [38, 39]. Taking into account these considerations, it could be though that significant and robust correlations could be obtained between the values of T_{leaf} and some gas-exchange parameters. However, crop physiological parameters such as stomatal conductance, transpiration or and net photosynthesis usually have a maximum point (under optimum conditions), meanwhile, the canopy or leaf temperature always changes with the environment, an hence, it shall not have a maximum or an optimum point. Because of this, it would be very difficult directly to establish the relationship of canopy temperature with those physiological parameters. This fact was described by Lima et al. [40], who concluded that the main problem of using the canopy or leaf temperature was related with the fact that T_{leaf} is highly dependent on climatic conditions (such as Tair, radiation, RH, DPV, etc). Even, many times the effects of these parameters in canopy temperature can be higher than the promoted effect by a water stress situation.

Attempting to this, and considering that as CWSI as I_g are highly dependent on the methodology applied, how can we decide which of them is the most profitable in terms of the crop-water status monitoring?. The response is not too easy, and up to day, many authors suggest that the main criteria to choose the best thermal index would be based on the study of the relationships between these indexes and crop water stress measurements taken at plant scale (that is, gas-exchange measurements or crop water potential). Thus, the knowledge of these relationships is required in order to apply this technique for a precise and robust estimation of crop-water status and irrigation scheduling. However, up to day, the main difficulty has been in the capability of obtain robust and universal relationships between the main thermal indexes and the most studied physiological variables. The causes are not very clear, although, some aspects such as the crop response variability to water stress under different edaphic and climate conditions, the different behavior of different varieties for a same crop, and the effects of climate conditions in leaf temperature (as it has been previously discussed), would be some of the reasons for not having been able to obtain robust relationships and universalize this technique for crop water monitoring.

Another question is the fact that there is not a normalization of measuring process. That is, thermal images can be taken from different ways; at plant, plot or field scale; and hence, the analysis is very different in function of the type of image. Moreover, the analytical process, is other aspect that would affect the final information obtained from the thermal image. As it has been previously discussed, data acquisition from the image is not a trivial aspect. It must be done with the highest precision and meticulousness, because, with a high probability, in the images will be find another parts of the plant (such as branches, stem, soil, sky, fruits, etc.) than have to be deleted and no considered for a good imaging.

Considering all these aspects is not easy (and maybe impossible) to define threshold values (or optimum ranges) in function of T_{leaf} or some thermal index for even crop (even more if we are talking about woody species). This fact does not occur when we are talking about other physiological variables, for instance the stem (or leaf) water potential, stomatal conductance or net photosynthesis. For all these reasons, the only way to obtain reliable data and relationships is to develop local calibrations, defining the main relationship of these thermal indexes with some physiological measurements.

CONCLUSION

Infrared thermography supposes an important advance in agriculture, particularly for those experiences associated with an improvement in the agricultural water resources management. Up to day, its great advantage is the robustness of the provided information, the speed with which these measurements are taken and the possibility of developing experiences at different scales. However, there are some limitations that must be overcome, some of them related with the high price of these systems, the need for highly qualified personnel, the need of data normalization throughout the use of some thermal indexes; the architecture of the crop the image analysis process.

Despite these limitations, future works should be oriented to improve this technique, being focused to standardize the methodology and image analysis. This works will allow improving the relationships between the stress indices and other more commonly used physiological variables, to make a decision with a greater strength.

ACKNOWLEDGMENTS

The author I. Garcia-Tejero has a contract co-financed by the Operational Programme of the European Social Fund (ESF) 2007-2013 "Andalucía is moving with Europe."

This work has been developed within the Regional Project 'INNOVASAR- Estrategias de riego deficitario integrando teledetección, modelización y experimentación en respuesta al cambio climático en Andalucía, PP.AVA.AVA201301.10.

REFERENCES

[1] García-Tejero, IF; Durán-Zuazo, VH; Muriel-Fernández, JL. (2014). Towards sustainable irrigated Mediterranean agriculture: implications for water conservation in semi-arid environments. *Water International.* Online first: http://dx.doi.org/10.1080/02508060.2014.931753.

[2] Chaves, MM; Zarrouk, O; Francisco, R; Costa, JM; Santos, T; Regalado, AP; Rodrigues, ML; Lopes, CM. (2010). Grapevine under deficit irrigation – hints from physiological and molecular data. *Annals of Botany* 105, 661-676.

[3] Costa, JM; Ortuño MF; Lopes CM; Chaves, MM. (2012). Grapevine varieties exhibiting differences in stomatal response to water deficit. *Functional Plant Biology*, 39, 179-189.

[4] Chaves, MM; Maroco, JP; Pereira, J. (2003). Understanding plant responses to drought – from genes to the whole plant. *Functional Plant Biology* 30, 239 – 264.

[5] Möller, M; Alchanatis, V; Cohen, Y; Meron, M; Tsipris, J; Naor, A; Ostrovsky, V; Sprintsin, M; Cohen, S. (2007). Use of thermal and visible imagery for estimating crop water status of irrigated grapevine. *J. Exp. Bot.* 58, 827–838.

[6] García-Tejero, IF; Durán-Zuazo, VH; Rodríguez, CR; Muriel-Fernández, JL. (2011). Water and Sustainable Agriculture. *Springer briefs in Agriculture*, 94 pp.

[7] Escalona, JM; Flexas, J; Medrano, H. (1999). Stomatal and non-stomatal limitations of photosynthesis under water stress in field-grown grapevines. *Australian Journal of Plant Physiology*, 26, 421-433.

[8] Costa, JM; Grant, OM; Chaves, MM. (2013). Thermography to explore plant-environment interactions. *J. Exp. Bot.* 64, 3937 – 3949.

[9] García-Tejero, IF; Muriel-Fernández, JL; Durán-Zuazo, VH; Arriaga, J. (2013). Uso eficiente del agua en la agricultura. *Instituto de Investigación y Formación Agraria y Pesquera. CAPDR*. 104 pp.

[10] García-Tejero, IF; Costa, JM; Lima, RSN; Durán-Zuazo, VH; Muriel, JL; Chaves, MM. (2014). Aplicaciones de la termografía de infrarrojos en la agricultura moderna. *Vida Rural*, 376, 60-66

[11] Jones, HG; Vaughan, RA. (2010). Remote sensing of vegetation: principles, techniques and applications. Oxford, UK, Oxford University Press.

[12] Jones, HG; Stoll, M; Santos, T; de Sousa, C; Chaves, M.M; Grant, OM. (2002). Use of infra-red thermography for monitoring stomatal closure in the field: application to the grapevine. *Journal of Experimental Botany*, 53, 2249-2260.

[13] Costa, JM; García-Tejero, IF; Chaves, MM. (2013). O uso da termografia na agricultura moderna. *APH*, 113, 30-34.

[14] Kaplan, H. (2007). Practical applications of infrared thermal sensing and imaging equipment. Tutorial Texts in Optical Engineering, Vol. TT75., 3rd ed. SPIE Press,Bellingham, Washington, USA, pp. 166.

[15] Fuentes, S; De Bei, R; Pech, J; Tyerman, S. (2012). Computational water stress indices obtained from thermal image analysis of grapevine canopies. *Irr. Sci.* 30, 523–536.

[16] Chaves, MM; Santos, TP; Souza, CR; Ortuño, MF; Rodrigues, ML; Lopes, CM; Maroco, JP; Pereira, JS. (2007). Deficit irrigation in

grapevine improves water-use efficiency while controlling vigour and production quality. *Ann. Appl. Biol.* 150, 237-252

[17] Pou, A; Diago, MP; Medrano, H; Baluja, J; Tardaguila, J. (2014). Validation of thermal indices for water stress status identification in grapevine. *Agricultural Water Management*, 134, 60-72.

[18] Zia, S; Spohrer, K; Merkt, N; Wenyong, D; Xiongkui, H; Müller, J. (2009). Non-invasivewater status detection in grapevine (Vitis vinifera L.) by thermograph. *Int. J. Agric. Biol. Eng.* 2, 46–54.

[19] Jones, HG; Serraj, R; Loveys, BR; Xiong, L; Wheaton, A; Price, AH. (2009). Thermal infrared imaging of crop canopies for the remote diagnosis and quantification of plant responses to water stress in the field. *Functional Plant Biology*, 36, 978-979.

[20] Grant, OM; Davies, MJ; James, CM; Johnson, AW; Leinonen, I; Simpson, DW. (2012) Thermal imaging and carbon isotope composition indicate variation amongst strawberry (Fragaria ananassa) cultivars in stomatal conductance and water use efficiency. *Environmental and Experimental Botany* 76, 7–15

[21] Jones, HG; Aikman, D; McBurney, TA. (1997). Improvements to infrared thermography for irrigation scheduling in humid climates. *Acta Horticulturae*, 449, 259-266.

[22] Maes, WH; Steppe, K. (2012). Estimating evapotranspiration and drought stress with ground based thermal remote sensing in agriculture: a review. *J Exp Bot* 63, 4671-4712.

[23] Grant, OM; Tronina, L; Jones, HG; Chaves, MM. (2007). Exploring thermal imaging variables for the detection of stress responses in grapevine under different irrigation regimes. *J. Exp. Bot.* 58, 815-825.

[24] García-Tejero, IF; Durán-Zuazo, VH; Muriel, JL; Jiménez-Bocanegra JA. (2011). Linking canopy temperature and trunk diameter fluctuations with other physiological water status tools for water stress management in citrus orchards. *Funct Plant Biol* 38, 106-117.

[25] Jiménez-Bello, MA; Ballester, C; Castel, JR; Intrigliolo, DS. (2011). Development and validation of an automatic termal imaging process for assessing plant water status. *Agricultural Water Management*, 98, 1497-1504.

[26] Zarco-Tejada, PJ; Berni, JAJ; Suárez, L; sepulcré-Cantó, G; Morales, F; Miller, JR. (2009). Imaging chlorophyll fluorescence with an airbone narrow-band multispectral camera for vegetation stress detection. *Remote Sensing Environ.* 113, 1262-1275.

[27] Wang, D; Gartung, J. (2010). Infrared canopy temperature of early-ripening peach trees under postharvest deficit irrigation. *Agric. Water Manage.* 97, 1787-1794.

[28] García-Tejero, I; Durán-Zuazo, VH; Arriaga, J; Hernández, A; Vélez, LM; Muriel-Fernández, JL. (2012). Approach to assess infrared thermal imaging of almond trees under water-stress conditions. *Fruits*, 67, 463-474

[29] Jones, HG. (1999). Use of infrared thermometry for estimation of stomatal conductance as a possible aid to irrigation scheduling. *Agricultural and Forest Meteorology*, 95, 139-149.

[30] Idso, SB; Jackson, RD; Pinter Jr, PJ; Reginato, RJ; Hatfield, JL. (1981). Normalizing the stress-degree-day parameter for environmental variability. *Agricultural Meteorology* 24, 45–55.

[31] Idso, SB. (1982). Non-water-stressed baselines: a key to measuring and interpreting plant water stress. *Agricultural Meteorology* 27, 59–70.

[32] Jackson, RD. (1982). Canopy temperature and crop water stress. *Advances in Irrigation Research* 1, 43–85.

[33] Jackson, RD; Kustas, WP; Choudhury, BJ. (1988). A re-examination of the crop water stress index. *Irrigation Science* 9, 309–317.

[34] Jones, HG. (1999). Use of thermography for quantitative studies of spatial and temporal variation of stomatal conductance over leaf surfaces. *Plant, Cell and Environment* 22, 1043–1055.

[35] Leinonen, I; Jones, HG. (2004). Combining thermal and visible imagery for estimating canopy temperature and identifying plant stress. *Journal of Experimental Botany* 55, 1423–1431.

[36] Meron, M; Tsipris, J; Charitt D. (2003). Remote mapping of crop water status to assess spatial variability of crop stress. In: Stafford J, Werner A, eds. Precision agriculture. *Proceedings of the 4th European Conference on Precision Agriculture*, Berlin, Germany. Wageningen: Academic Publishers, 405–410.

[37] Irmak, S; Haman, DZ; Bastug, R. (2000). Determination of crop water stress index for irrigation timing and yield estimation of corn. *Agronomy Journal* 92, 1221–1227.

[38] Fuchs, M. (1990). Infrared measurement of canopy temperature and detection of plant water stress. *Theoretical and Applied Climatology* 42, 253–261.

[39] Jones, HG. (1992). Plants and microclimate, 2nd edn. Cambridge: Cambridge University Press.

[40] Lima, RSN; García-Tejero, I; Lopes, TS; Costa, JM; Vaz, M; Durán-Zuazo, VH; Chaves, M; Glenn, DM; Campostrini, E. (2015). Linking thermal imaging to physiological indicators in Carica papaya L. under different watering regimes. *Agric. Water Manage.* In revision.

In: Arid and Semi-Arid Environments
Editor: Gabriel Ferguson
ISBN: 978-1-63483-383-7

Chapter 3

GEOMORPHODIVERSITY OF THE EJINA BASIN IN INNER MONGOLIA, CENTRAL ASIA: IMPLICATIONS FOR LATE QUATERNARY LANDSCAPE EVOLUTION AND PALAEOENVIRONMENTAL CHANGE

Bingqi Zhu[1], Jingjie Yu[1], Patrick Rioual[2], Yan Gao[3], Yichi Zhang[1], Leilei Min[1], Chaoyang Du[1] and Heigang Xiong[4]

[1]Key Laboratory of Water Cycle and Related Land Surface Processes, Institute of Geographic Sciences and Natural Resources Research, Chinese Academy of Sciences, Beijing, China
[2]Key Laboratory of Cenozoic Geology and Environment, Institute of Geology and Geophysics, Chinese Academy of Sciences, Beijing, China
[3]Centro de Investigaciones en Geografia Ambiental, Universidad Nacional Autonoma de Mexico, Campus Morelia, Mexico City, Mexico
[4]Key Laboratory of Oasis Ecology, Ministry of Education, Urumqi, China

ABSTRACT

The north-south trending basin of Ejina in Inner Mongolia of northernwestern China, Central Asia is located at the lower reaches of the Heihe River watersheds and is a regional tectonic depression limited to the west part of the Alashan Plateau. Though neotectonic activity is believed to be the major factor in the evolution of the topography, detailed documentation and analysis of climatic landscape features like alluvial fans, desert plains, aeolian dunes and bedrock gorges and their environmental implications are lacking. The present study is a site-specific documentation of landforms developed in the wide part of the Ejina Basin, with the aim to identify the climatic landforms based on the method of climatic geomorphology and to evaluate its landscape evolution and response to palaeoclimate changes. The morphodynamics of older landscapes are recognized by making comparison with the present climate and its corresponding landscapes. Occurrence of desert plains and aeolian dunes in the central basin is inferred to be the products of modern hyper-arid climate. No active geomorphoclimatic dynamics for development of desert bedrock gorges and pediments is observed at the margins in the present day conditions, thus we infer an obvious linkage of these landforms to the humid phases, which provided high energy runoff for the formation of landforms associated erosional features. Clear evidences testifying the basin-scale shifting and transformation of different morphoclimatic zones in the Ejina basin are observed, which prove that the main geomorphic unit is changed from an alluvial-lacustrine plain to a desert plain. Pediments and desert gorges are both relict landforms associated with more humid and colder climate during the last glacial period. However, the coexistence of diverse landscapes and the consequent geomorphodiversity in the basin should be a compound result of surficial processes other than glaciations. Fluvial/alluvial actions, aeolian, abrasion/deflation, aggradation and degradation processes together with climatic changes have conjunctionally played important role. The climate and hydrological conditions of the basin during the last glaciation and during the Early Holocene were much better than at present, possibly having an average annual precipitation ranged between 60~350 mm on the margins of the basin and being characterized in the central basin by high lake levels during ca. 39-23 ka BP but great fluctuations during Holocene. Because the Asian summer monsoon was quite weak during the glacial times, the periods of lower aridity during the late Pleistocene in the Ejina Bain could be induced by an increase of the westerlies and a weakening of the Asian winter monsoon on the arid areas of the central Asia.

Keywords: Desert basin, landscape patterns, geomorpholodiversity, late quaternary climate, ejina, Central Asian

1. Introduction

Northern China in Central Asia exhibits distinct variations in climate and hydrology during late Quaternary resulting in variation in the form and size of landscapes (Yang et al., 2010, 2011a, b). Due to the absence of long and continuous sedimentary sequences in desert area, finding effective substitute in the deserts of world to understanding the Quaternary climatic history is still a difficult issue (Yang 2001; Yang et al., 2004). Knowledge of the late Quaternary climatic history of northwestern China is crucial to understanding the evolution of the East-Asian climate system (Liu et al., 1996, 2002; Liu and Ding, 1998; Sun et al., 1998; Feng et al., 1998; Yang X et al., 2004, 2011b, 2012; Ding and Chan, 2005; Herzschuh, 2006; Wunnemann et al., 2007; Stevens et al., 2008; Yang S et al., 2008; Mason et al., 2009; Lu et al., 2011; Li et al., 2014).

Table 1. A concise state of major landscapes occurred in arid zones and their related climatic characteristics (Hoevermann, 1985)

Landscape division	Geomorphic feature	Climatic characteristics
Glacial action zone	Glacier, glacial erosion and glacial deposition morphology	Mean annual temperature of -1~-13°C (varied with arid index)
Valley landscape a) Forest area b) Steppe zone	Weak geomorphodynamic, soil formed, seasonal rivers, headward erosion, humus layers development, ephemeral rivers, mild water erosion, loess-like deposition, calcareous soil	Annual mean temperature of 1~-5°C the coldest month mean temperature of <-2°C
Pediment	Conical-shape deplanation and accumulation patterns, radial rill networks	Semi-arid (precipitation 150~350 mm/a), strongly freeze-thaw weathering
Desert Gorges	Water-eroded patterns with steep wall-sides	Arid (precipitation 60-150 mm/a)
Desert Plain	Gently inclined plains covered by sandy surface (formed jointly by wind and water)	Arid (precipitation 30-50 mm/a)
Aerodynamic topography	Dunes, wind-eroded deflation depressions and Yardangs (aeolian patterns)	Arid (precipitation <30 mm/a)

Until now, acquired clues reflecting desert palaeoenvironmental history in north China are on a large degree derived from proof of loess and palaeosol sedimentary sequences located at the peripheric or remote areas of the deserts (e.g., Liu and Ding, 1998; Guo et al., 2002; Sun, 2002). Direct index derived from local desert areas are seldom used to evaluate the Quaternary changes of desert environment in China.

The landscape conception in the sense of climatic geomorphology seemed to be a promising method to compensate for this deficiency. As an earth science opinion, for example the Davies's theory of geomorphology, the topography of the Earth's surface is the result of the balance between endogenic and exogenic forces. Endogenic forces trigger vertical movements that generate large mountain belts and depressions, whereas exogenic forces work progressively to denude the resulting relief. This permanent interplay of forces has been performed at different scales on the Earth's surface during the whole of geological history. Climatic geomorphology are defined as the discipline that identifies climatic factors such as the intensity, frequency and duration of precipitation, frost intensity, direction and power of wind, and it explains the development of landscapes under different climatic conditions (Ahnert, 1996). For example, based on consideration of two climatic parameters (mean annual temperature and total annual precipitation), Peltier (1950) established their relationship with five key geomorphic processes (chemical weathering, frost action, pluvial erosion, mass movement and wind action) and distinguished two different morphogenetic elements (weathering processes and the transport agents of the resultant materials). In this perspective nine different morphogenetic regions were postulated (Peltier, 1950). Taking climate as a main dynamic of landscape formation, Hoevermann (1985) established a corresponding relation between landforms and climate types in arid areas, as show in Table 1. Pope et al. (1995) proposed a conceptual model for understanding landscape evolution and geographical variations in weathering. From the study of micro-landform development, we can obtain valuable information for a better understanding of the long sequence of events that has resulted in present-day landforms (Gutierrez, 2005). So the landscape conception in the sense of climatic geomorphology seemed effective to bridge the gap that finding effective substitute in the deserts of world to understanding the past environmental change.

The concept "geomorphodiversity" borrows from the analogue and well known biodiversity, and especially, also the less well known geodiverisity (Gordon et al., 2012). Based on the two sciences, recently a new paradigm for climate science, climatediversity, is proposed (Beggs, 2013). Due to the quick progress of the earth system science, the geological and climatological communities are now actively emphasising on an integrated science of biodiversity, geodiverisity and climatediveristy (Beggs, 2013), in which the geomorphodiveristy is taken as an important component. As defined by Panizza (2009) and Testa et al. (2013), geomorphodiversity is a critical evaluation of the geomorphological characteristics and dynamics of a territory. Whatever at the macroscale or global scale, the implication of both observed past and projected future climate and environmental change for geomorphodiversity is a research question in need of great attention. This work is the first time attempt dedicated to the geomorphodiversity study on the desert basin landscape of northern China.

Ejina Basin lies on the west part of the Alashan Highland and at the central part of the desert belt in north China (Figure 1a). The geographical location of the basin is in one side on the northern fringe of the Asian summer monsoon and in the other side on the geographical latitude of westerlies (Figure 1b). So the Ejina Basin is a potentially significant place for the research on global climate change and for the studies about regional responses to global climate change. On the other hand, the Alashan Plateau had been an old stable region but became active since Mesozoic era (Guo et al., 2000). The environmental evolution of this area was affected by each tectonic phase caused by reciprocal compression between Eurasian plate, Indian plate and pacific plate, so geomorphological and paleoenvironmental research in this area also provide evidence for studying environmental impact of the upwelling of the Qinhai-Tibet Plateau and formation of East Asian monsoon (Guo et al., 2000). The current climate in the Alashan Plateau including the Ejina Basin is hyper-arid and the current hydrological setting are single, with only one river (the Ruoshui River) flowing into the Ejina Basin (Zhu et al., 2014a). Outside the Ejina, no fluvial processes occur in the Alashan Plateau (Zhu and Yu, 2014). It means that the climate and hydrology at present in Ejina is relatively monotone and stable. While, field investigations show that the regional forms of existing landscape in the Ejina is diverse (Zhu et al., 2014b), evidently a result of multiple geomorphlogical processes about periglacial, fluvial and alluvial action, aggradation and degradational processes. Besides, there have lots of active dunes occur along the Ruoshui river channel in the basin (Zhu et al., 2014a). The dunes are surprisingly much lower than those in the neighboring area (the Badanjilin Desert) in the Alashan Highland (Zhu et al., 2014a). As widely accepted, dune formation depends mainly on arid climate (e.g., wind intensities) and the availability of sand (Yang et al., 2011a). Therefore the dunes and the landscapes in and around the Ejina Basin are worthy of special examinations. Until now, however, documents in literatures concerning about the landscape and palaeoenvironmental evolutions of the Ejina Basin are still little. On the basis of examining landscapes distribution and formation, this paper intends to provide an insight into the geomorphological evolution and related late Quaternary variation of the climate in the Ejina Basin.

2. Regional Setting and Methods

The Ejina Basin under study is located in the west of the Alashan Plateau, between 40-43°N and 99-102° E of the middle-latitude zone of Northern Hemisphere (Figure 2). The Alashan Plateau lies in the western Inner Mongolic, NW China, and consists mainly of the gravelly plains of the Alashan Gobi together with the big sandy areas of the Badanjilin and the Tenggeli Sand Seas (Figure 2a). The Qilian Mountains, Tibetan Plateau and the highlands of the Beishan, Mazongshan, Gobi Altai and Helanshan form the natural boundaries of the Alashan Plateau to the south, west, north and east (Figure 2a). It has an area of 300,000 km^2 in altitude between 900 and 1300 m a.s.l. (average 1,000 m asl), and a relief that depresses gradually from east to west and from south to north. Gobi, sandy desert and relict mountain are the major landforms (Figure 3). Located in the inland of northwestern China and adjacent to Mongolia High-pressure center, Alashan Plateau is one of the driest areas of the world, characterized by extreme continental temperate climate conditions with hot summers and cold winters. Precipitation derived mainly from westerlies and the Asian summer monsoon shows a strong interannual variability. Drainage system is poorly developed in this plateau, except the west part where the Ruoshui River is a big seasonal inland river with running water from July to October every year.

The Ejina basin is north-south trending in topography and is a regional tectonic depression limited to the east by the Badanjilin desert units and to the east by Mazongshan Mountain Units (Figure 2). Surface drainage in the Ejina is constituted by the Ruoshui (Ejina) River and its ephemeral tributaries, among which the most important are the East River and West River both coming from the Heihe River, one of the three largest inland river systems in China. The elevation of the East river course declines from Langxinshan to Lake Sugunuoer about 160 m (Figure 4a). The climate in the basin is arid under desert conditions. The harshness of the present climate is manifested by wide annual temperature variations (77°C), atmospheric transparency, low humidity, winter snow precipitation and very scarce pluvial precipitation (Figure 4b-d).

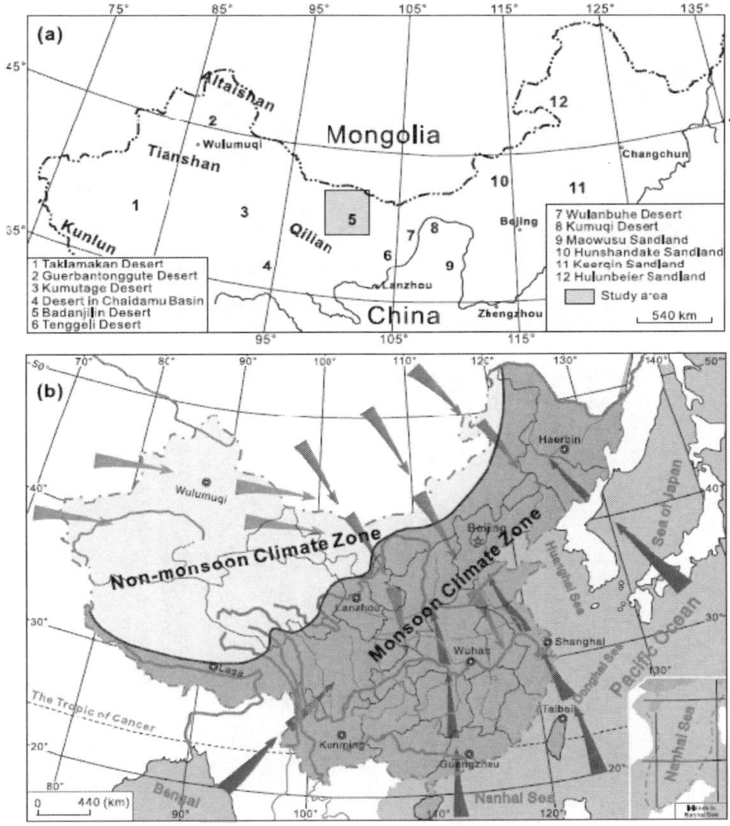

Figure 1. Distribution of the deserts in China and the location of the study area (a) and the distribution of monsoon climate zone and non-monsoon climate zone (westerly zone + cold-arid regions in the Tibetan Plateau) in China (b) (Cited from Zhu et al., 2014b).

The mean annual temperature is 8.8°C at Ejina during the last five decades (Figure 4d), with a maximum daily temperature of 41°C (July) and a minimum of -36.4°C (January). The rainfall regime is continental, with summer rains (Figure 4b) and only with very low average frequency of days with rain.

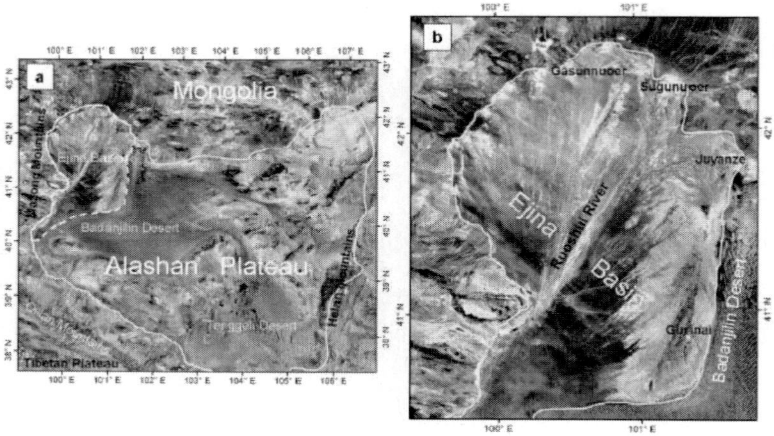

Figure 2. Overview remote-sensing images (Landsat TM, RGB 7-4-2) of the Alashan Plateau (a) and the Ejina Basin (b) (Cited from Zhu et al., 2014b).

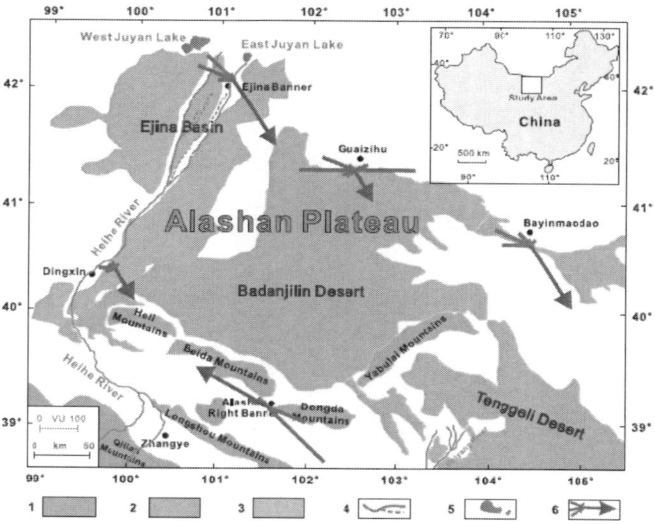

Figure 3. Landscapes of the Alashan Plateau. Legend: 1 sand seas, 2 Alluvial and fluvial sediments associated with the Heihe (Ruoshui) River, 3 Adjacent mountain ranges, 4 Rivers, 5 Lakes, and 6 sand roses (Fryberger and Dean, 1979) for each of the five surrounding weather stations, with aubergine lines showing winds capable of transporting sand from various directions (DP) and red lines with arrows indicating the resultant sand transport trend (RDP) (modified from Yang et al., 2011a).

The average annual precipitation in the last five decals recorded in the weather stations of Ejina is 35.6 mm (Figure 4c). From a climatic viewpoint, this area can be classified as hyper-arid (120 > p> 60 mm) or eremitic (60 >p > 30 mm), depending on the subzone (Perucca and Martos, 2012). Prevailing winds come from the west (Figure 3), and the northwest winds are almost constant occurrences during August-September.

Characterization of local and regional landscapes in the Ejina desert basin is described mainly relying on the LANDSAT ETM+ data, topographic maps and field examinations. The identification and classification of landscapes is to a considerable extent based on the system of climatic geomorphology, as show in Table 1. The field investigation in the Ejina area concentrated on the upper, middle and lower reaches of the Ejina River that flows from the Langxinshan northwards into the Lake Sugunuoer (East Juyanhai). Detailed fieldwork was carried out in the whole drainage region. Chinese 1:100,000-scale topographic maps, 1: 500,000-scale Google-Earth images and GPS were used for the orientation in the field.

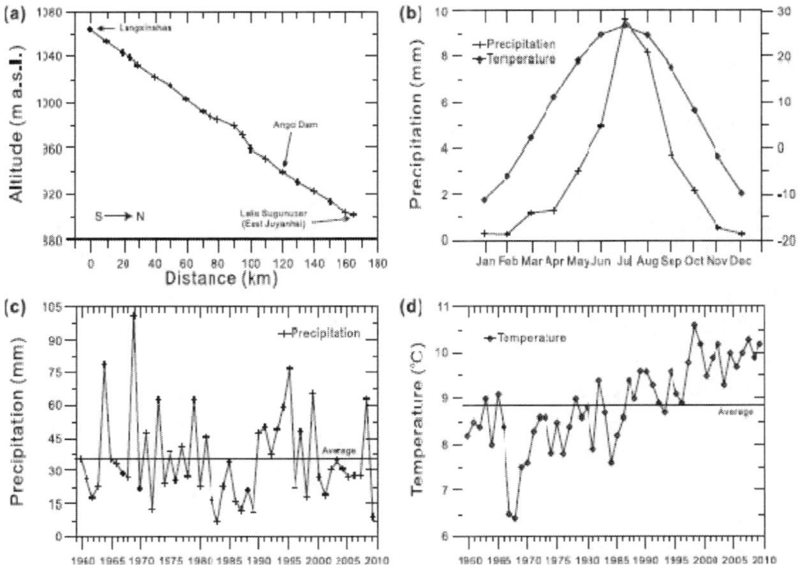

Figure 4. A topographic profile of the S-N trending line (along the East river from Langxinshan to Lake Sugunuoer), the seasonal distribution of average precipitation and temperature (50 years during 1960-2009) (b), variations of the average annual precipitations (c) and temperatures (d) (50 years during 1960-2009) in the Ejina Basin.

3. Results and Discussion

3.1. Identification of Dominant Landforms in the Ejina Basin

The geomorphological conditions of the Ejina Basin are complexity. It is a privileged place to observe the relationship between mountain, peneplain and basin playas which are directly connected to the genesis of the Ejina landscapes (Zhu and Yu, 2014; Zhu et al., 2014a). The Ejina geomorphological complexity is firstly determined by lithology and tectonics factors, which are obvious morphotectodynamic processes indicated by significant gradients between mountain peaks (Mazong Mountains) and basin floors (Juyan Lake) in the Ejina (Zhu et al., 2014b). Besides tectodynamics, satellite imagery and field observations also show that the most widely spread landscapes in the arid regions of China are desert plains and aerodynamic landforms (Hoevermann 1985; Yang 2001; Yang et al. 2004, 2011a, b), of which the Ejina Basin is the most representative (zhu et al., 2014b). Detailed identification and explanation about the types of the dominant landforms in the Ejina Basin are listed below.

3.1.1. Desert Plains

Tectonic activity often intensifies surface erosion to produce a large amount of debris, which are transported by rivers and/or glaciers and deposited on mountain piedmont and basin depression, forming vast flat alluvial fans. Under dry conditions, strong winds blow away the silt and sand fractions of sediments, and leave a gravel cover on the alluvial fan surface, thus forming the desert plain. Quite a large part of areas in the Ejina Basin is typical for desert plain (Figure 5). The geomorphological process of desert plain formation can be divided into erosion and accumulation. The accumulation plain dominates in the central part of the Eijina basin and in areas close to the aerodynamic relief. In the transitional zone to the surrounding highlands, erosion plains prevails (Figure 5a). In the Ejina Basin there are depositional desert plains that are called "Gobi" in Chinese terminology. The sedimentary strata of the desert plains are mainly proluvial-alluvial deposits (Figure 6a), complicated sequences with sand, gravel, silt and clay sediments involved alluvial, fluvial, lacustrine and aeolian processes (Figure 6b), thick clayey deposits by lacustrine processes (Figure 6c). Slightly eroded shallow sections by fluvial processes are often observed in desert plains (Figure 6d) and huge lacustrine profiles occur in the lower northern part of the desert plain (Figure 6e).

Figure 5. Landscape of desert plains and typical reg of clasts developed on the alluvial plain in the Ejina Basin.

Figure 6. Sedimentary Profiles in the Ejina Basin: (a) proluvial-alluvial deposits (in desert plain), (b) complicated sedimentary sequence including sand, gravel, silt and clay sediments involved alluvial, fluvial, lacustrine and aeolian processes (in desert plain), (c) thick clayey deposits, (d) erosional terrace of river bank (margin of the basin), (e) palaeo-lacustrine section (at the low terrain of the north basin).

The upper sediments of Gobi desert are sand, silt and gravel in compositions and the gravel lie mostly in a relatively thin layer above the surface, shown as desert pavements (Figure 5b, d). Desert pavement can originate via a single or various processes, which indicates a problem of convergence of forms (McFadden et al., 1987). Fundamentally the processes can be extrinsic in character, produced by water and aeolian action, and intrinsic, in which the interior of the soil is modified by differentiation processes. The most universally applied process of pavement genesis is deflation, whereby fine particles are exported by the wind, leaving a residue of coarse material that goes on to constitute the pavement.

3.1.2. Pediment

In the system of climatic geomorphology, the pediment represents a conical relief characterized by a channel net. Frost weathering producing loose rock materials during the winter is crucial to the formation. Water from rainfall in the summer brings these materials to the streams and alluvial fans are consequently developed. Like desert plains, pediments can be distinguished into erosive and depositional sub-forms. Around the western and northern edges of the Ejina Basin, pediments appear on the foot zone of mountains and in the higher sections of desert plains (Figure 7).

3.1.3. Desert Gorges

The desert gorges are widely distributed in the surrounding mountain slopes like the Mazongshan Mountains in the western of the Ejina Basin (Figure 8). The desert gorge storey is present in the upper part of pediments and the desert plains and the gorges usually dissect the mountain fronts. The landscape of gorges was produced by water and there is evidence of former severe erosion (Figure 8a-c). In some places, loose sand layer begin to cover the gorge surface by aeolian and desiccation-deflation processes, owing to lack of water (Figure 8d).

3.1.4. Aeolian Landform

The aerodynamic landform refers to the actual wind-sculpted relief including dune fields of various forms and wind-eroded rocks like yardangs and deflation depressions. In Chinese terminology, desert is always related with dune fields (Yang et al., 2004). The dune landscapes in the Ejina Basin are mainly located at areas along the two bank sides of the Ejina River (Figure 9) and surrounded by Gobi and desert plain (Figure 10).

Figure 7. Landscape of pediment in the Ejina Basin.

Figure 8. Landscape of desert gorges in the Ejina Basin.

Based on vegetation conditions the dunes can be divided into three main types: active, semi-active and fixed dune (Figure 10). On the dune land surface *Populus euphratica, Chionese tamarisk, Artemisia, Agriophyllum arenarium* and other plants are scarcely distributed (Figure 9c-d). The height of 23 dunes examined by us in field ranges from 40 to 1.5m, with an average 11 m (Zhu et al., 2014a).

Figure 9. Aeolian landform (dune fields) in the Ejina Basin. (a) Single dunes and dune chanins, (b) sandy sheets, (c) and (d) Dunes accompanied by Tugai forests (Populus euphratica) and shrub vegetations (e.g., Chionese tamarisk) along the Ejina river banks.

These figures confirm that the height of dunes changes regularly on a local scale but the dune types do not differ greatly from place to place (Zhu et al., 2014a). The three types of dunes occur at any locations. Compared with the neighboring dune filed in the Badanjilin Desert, where the height of most dunes ranges from 333 to 185m, the dunes in the Ejina are much lower and relatively younger at their early stage of dune development (Zhu and Yu, 2014). Differing from the dune shapes in the Badanjilin Desert, where complex megadunes predominate and secondary dunes of various forms are overlapped on the windward slope of these megadunes, the dunes in the Ejina are almost single in figuration and are mainly dune chains, transverse and net dunes (Figure 9a-b). The ridges of these dunes are not straight and both lee and windward slopes are gentle (Figure 9a-b), with slope angles less than 5°. In this sense, three types of aeolian accumulation can be classified: sand ripples, Draas and dunes (Busche, 1998). On the slope of most of the dunes in Ejina there are no any calcareous layers, clearly different from the dunes in

Badanjilin, indicating that most of the dunes in Ejina Basin are under early stages of dune development and no former configuration of the dunes. Compared with our field investigations in other deserts before, pyramid or star dunes occur widely in the Taklamakan and the Badanjilin deserts, the largest and the third largest sandy desert in China, respectively (Zhu and Yu, 2014). Observations in the Taklamakan show that some pyramid dunes have been created by erosive processes and in the Badanjilin the multi-directional winds presumably played an important role (Yang et al., 2004). While in the Ejina Basin, this kind of dunes is rarely observed. A seasonal change of directions of prevailing winds is occurred in the records of Ejina weather stations (Figure 3) but no balanced wind systems in different directions and the wind energy is weaker than that in the neighboring Badanjilin. This is maybe an important major reason that no clear pyramid-type of dunes develops in the Ejina.

Figure 10. Active dunes developed on the Gobi deserts.

3.1.5. Fluvial Landforms

The fluvial landform is only existed in the Ejina Basin on the Alashan Plateau (Zhu et al., 2014b). Ruoshui (or Ejina) River, as the low reaches of the Heihe River, is the unique river flowing through the Ejina Basin.

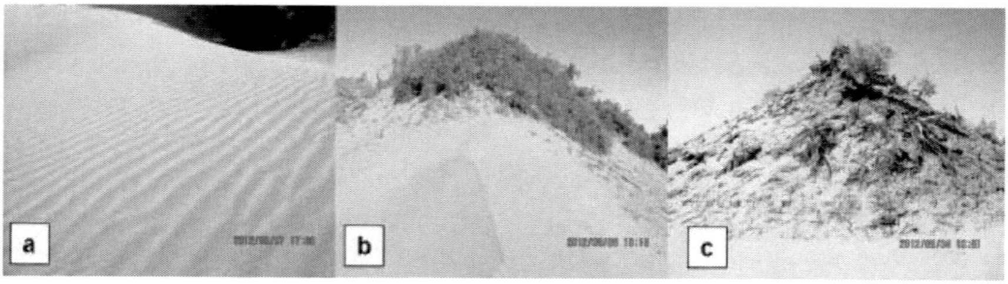

Figure 11. Active dune (a), semi-active dune (b) and fixed dune (c) in the Ejina Basin.

Figure 12. Fluvial landscapes in the Ejina Basin. (a) A satellite image of the river channel (East River), gobi and dune distribution on a short section in the central part of the Ejina basin (form Google Earth), (b) wide and flat river channel (no water flowing by at the time), (c) sandy sedimentation in the riverbed, (d) intermittent flood discharge with muddy water derived from the Qilian Mountains.

The river emerges from the south corner (Langxinshan) of the Ejina Basin and follows a S-N directional course of ca. 160 km before flowing into the Lake East Juyanhai (Figure 4a) at the northmost part of the basin. Due to the flat and wide terrain (Figure 12a-b) the relief of Ruoshui watershed is generally much smaller than that of watersheds in other parts of the Heihe River. The fluvial landforms are characterized by low-gradient courses and small slopes (Figure 12b), with a modal angle (0.06°) less than 0.1°. The stem river channel of the East River is about half to 1 km in width (Figure 12a-b) and the river beds are characterized by sandy sediments (Figure 12b-c).

Water supply of the Ruoshui River is markedly intermittent flood discharges with muddy water (Figure 12d), which was derived from the upland water in the Qilian Mountains areas during the spring and rainy seasons. The duration of floods is short recently because of water conservancy programs for the Heihe System at the middle reaches. While sediment yields in Ejina is highly significant, especially in the Quaternary history. Except for exists of the modern river channels, lots of the palaeo-river channels still can be identified both from the morphology in the field and from historical maps and satellite images (Figure 2a).

3.1.6. Modern and Ancient Lakebeds

In the low terrains and distal areas of the alluvial fans in the Ejian Basin (mainly the north part), modern and ancient lakebeds are widely distributed (Figure 13). There are also extensive areas with lacustrine and paludal sediments distributed at the east, central and northern peripheries of the Ejina Basin. The Gurinai Lake basin extends 120 km from south to north and about 50 km from east to west. The Garshunnuoer (Figure 13b) and Sugunuoer lake (Figure 13c) areas are about 100 km long in the east-west direction and up to 40 km wide from south to north. The maximum dimension of the Guaizihu Lake and Juyanze basins is about 60×60 km. These ancient lake basins have now thoroughly dried up in the natural environment, and the area of recent day playas, which are assumed to show evidence of higher lake levels in the past or even the existence of a palaeolake (Wunnemann et al., 2007; Hartmann and Wunnemann, 2009), cover an area of about 15,000 km^2. Viewing from remote-sensing images, these lakes were apparently the terminal lakes of the Roushui River (Figure 13). Fluvial sediments alternate with the lacustrine deposits in many profiles (Figure 6b). In some areas of the former lakebeds there are still whiter-color clayey sediment strata (Figure 6e). Along the Ejina River (with two tributaries, East River and West River) course in the central of the Eijian Basin there are thick layers of lake sediments that were deposited in previously strong hydrological conditions (Figure 6c).

Figure 13. Overview remote-sensing images (Landsat TM, RGB 7-4-2) of the north basin (a), the Lake Garsunnuoer (b), the Lake Sugunuoer (b) and the Lake Juyanze (d) in the Ejina Basin.

3.2. Geomorphodiversity of the Ejina Basin

3.2.1. Geomorphological Differentiation in a Desert Environment

Geomorphologists (e.g. Meigs, 1953) have defined deserts as a cold deserts when the prevailing mean temperature of the coolest month is less than 0°C. Accordingly, the Ejina Basin is under a cold desert environment because ground surface temperatures in the area have been registered to vary between 25°C and –15°C. The most significant characteristic of this environment is the great thermal annual oscillation that occurs. Due to its latitudinal position and to the anticyclonal domain, the Ejina Basin is characterized by great insolation and the air temperature can reach considerable values in summer, as much as 40°C. The warm winds in the desert regions cause the desiccation of many areas. However, the geomorphic criteria that can be used to establish classifications of arid zones can vary a great deal. The application of rules about dominant processes and general morphological types in desert environments allows us to establish geomorphological differentiations. Two important dominions can be recognized in deserts as a function of the main acting processes (Cook et al., 1993). One is the aeolian dominion and another is the fluvial dominion (Cook et al., 1993). Both dominions can be superimposed in the study area, because the activity of the aeolian and fluvial processes coexists in wide spatiotemporal scale in the Ejina Basin, as two examples can be seen from the basin sedimentary record with interbedded aeolian and fluvial sequences (Figure 6b) and the distribution of dunes along the river banks (Figure 9). On the other hand, it indicates that the climatic changes that have must taken place during the Quaternary have produced important modifications in the aridity of the basin. As a consequence, the activity and intensity of aeolian and fluvial processes seems to be modified considerably.

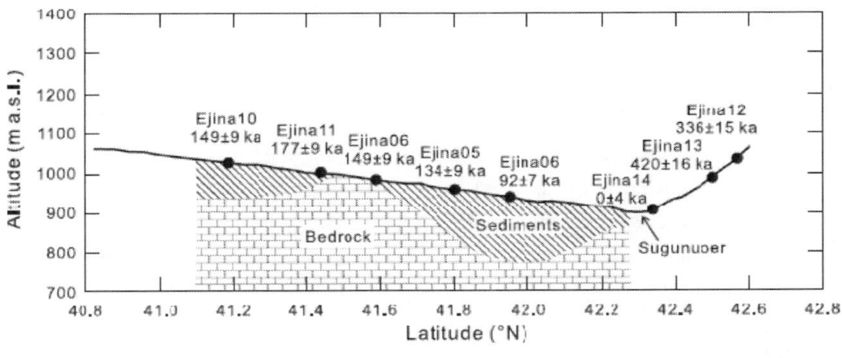

Figure 14. Variations in altitude, sediment thichness (Zhang et al., 2006), and exposure ages of the quartz gravels (Lv et al., 2010) from the Gobi deserts along a SW-NE transaction in the Ejina Basin.

The application of the geomorphic criterion in desert, based on the type of dominant morphologies and in the geologic structure, along with the degree of tectonic stability, permits distinguishing between shield/platform deserts and mountain/basin deserts (Mabbutt, 1977). Shield/platform deserts are formed fundamentally from cratonic areas preformed basically by Precambrian rocks and platforms of later age. Their main characteristic is the flat relief, interrupted by recent volcanic massifs and by important fractures. The mountain/basin deserts are constituted by long mountainous alignments separated by depressed areas. Evidently, the Ejina desert belongs to mountain/basin desert. Its marginal relief in a number of highland places was shaped during the alpine orogeny in the Mesozoic and Cenozoic ages. As a consequence of the topographical contrast the high areas have suffered a constant denudation and the resultant materials have lodged in the depressions, in the form of coalescent alluvial fans (Figure 2). In the distal areas of the alluvial fans the run-off waters are collected in closed depressions that form ephemeral lakes (Figures 2, 13) in which takes place an evaporitic sedimentation (playa-lake).

3.2.2. Geomorphological Inconsistency of Landforms in the Ejina Basin

Formations of desert gorge and pediment are both related with frost weathering processes under cold and wet climate (Hoevermann, 1985). Comparing the landscape type with the present dominant morphodynamic process and climate, e.g., desiccation-denudation processes and temperate-zone climate, implies that there is an inconsistency in the extensions of pediments and desert plains in the areas of the Ejina Basin.

The climatic precondition for the formation of a pediment is a cold winter and an average annual precipitation of 150-300 mm (Hoevermann, 1985). Frost weathering processes in winter and alluvial processes in summer is crucial to the formation of pediment. However, the average annual precipitation and temperature at present is only ca 35 mm and 8.8°C in the Ejina Basin, respectively (Figure 4b). The desiccation-denudation processes are predominated in a full year and the alluvial processes have decreased considerably under present climatic conditions. So an actual formation of pediment does not take place in the Ejina Basin under present climatic environment. Strong erosion processes by water are needed for the formation of desert gorge landscape. The critical amount of annual precipitation for developing desert gorges is thought to be 60-150 mm (Hovermann, 1985). At present these gorges areas are often bare and present frost weathering processes are quite weak in the Ejina Basin. Similar to the pediments and alluvial fans, desert gorge formation in the Ejina basin is now inactive owing to the lack of water. According to Hovermann (1985), whatever desert gorge and pediment landscapes, their formations both reflect much more annual mean precipitation and lower temperature than present (Table 1). Thus the pediments and desert gorges surrounding the Ejina Basin are relict landforms associated with more humid and colder climate in the past.

On the other hand, small shifting dunes are distributed widely along the Ruoshui river banks (Figure 9). According to Hovermann (1985), typical aerodynamic relief often occurs in the area with an average annual precipitation of less than 30 mm and the dry climate promotes the development of shifting dunes, which become stabilized under moister conditions. To some extent, river channel and the nearby areas are somewhat wetter conditions, so the occurrence of shifting dunes along the river sides (but not far away from there) should be an abnormality. Sand source/supply and the role of retention effect by vegetations (such as *Populus euphratica*), as well as unenen ground (due to shrub vegetation) and weak deflation processes nearby river may be significant factors for the formations of these dunes other than arid climate.

3.2.3. Geomorphological Transformation of Landforms in the Ejina Basin

Filed Investigations and sedimentary and mapping analysis suggest that the landscapes and the boundary of different landscape types in the Ejina Basin have undergone changes or transformation. At present the former pediments around the Ejina Basin are undergoing transformation towards desert plains in the lower terrains and towards desert gorges in the higher locations. In many locations the slopes with desert gorges are being buried under aeolian sand (Figure 8d). The gorges landscape is thus being transformed to an aerodynamic landforms assemblage.

In the marginal areas of dune fields, thin layer of gravel with a maximal diameter of 3-5 mm covers ground surface (Figures5d and 10). This occurrence of gravel layer is indicative of former more powerful winds effecting on the berried alluvial/pluvial sediments. The dunes generally possess small geometric parameters and sit on the surface of earlier desert plain landscapes, meaning that the present-day active shifting dunes are younger and distributed on an old relief surfaces. In genral, the development of shifting dunes is caused by climate conditions with intensive wind and shortage of precipitation. Its formation represents a transportation and collection of unstable/loose sandy materials and is related with an increased aridity index (Hoevermann, 1985). According to the conceptions of climatic geomorphology, the morphodynamic process and sand transportation for dune accumulation takes place in case the wind becomes stronger and the availability of water decreases. Such kind of shifting does not cause a change of landscape, but it is an evidence for regional climatic aridification and land desertification.

Desert plain, as a main geomorphological unit in the central Ejina Basin, was transformed from alluvial fan by alluvial depositions of the Ruoshui River. This can be proved by a wide exist of basin sedimentary stratigraphy with thick alluvial sedimentary strata buried under Gobi surface (Figure 6). The alluvial fan in Ejina had extended northward about 160 km and changed from alluvial plain to alluvial/lacustrine plain, and then to lacustrine plain. In terms of climatic geomorphology, desert plain refers to a relief that undergoes a morphodynamic process with balanced wind and water dynamics (Hoevermann, 1985). The relief energy on the desert plains is relatively low and the ideal amount of average annual precipitation for forming desert plains is 30-60 mm (Table 1). Viewing from this opinion, the current climate in Ejina is suitable for the formation of desert plain. However, the present-day desert plain in Ejina is still in the beginning stage of desert plain formation, because the shape of large-scale alluvial fan is still recognizable from the remote-sensing images (Figure 2) and there are also lots of relicts of riverbeds and fluvial terraces. Owing to the presence of groundwater and river used for irrigation, oases are developed in this zone. Along the riverbanks of the Ruoshui River the original desert plain has been replaced by an agricultural landscape, indicating a strong influence of human activities on regional landscape evolution. On the peripheral edge of the Ejina Basin, field investigations confirm that former alluvial landforms like channels are still well preserved (Figure 5a). During years of richer precipitation, the vegetation of the desert steppe is well developed (Figure 5c). The plant communities consist mainly of shrubs and half-shrubs. Therefore the landscape here can be seen as a transitional form between the desert plain and the pediment.

The Geomorphological transformation in the basin can be seen as a general increasing tendency of the climatic aridity. Several ca 3-4 m deep profiles located in the central part of the Ejina Basin consists of the following materials from bottom to top: a) interbedding of fine aeolian sand and coarse fluvial deposits; b) lake sediments; c) interbedding of aeolian sands and marsh deposits and d) interbedding of aeolian sands and sandy silts (Figure 6). This sequence leads to the interpretation that the landscape during the earliest period was a desert plain because of the occurrence of aeolian and fluvial deposits on the lowest part. Thereafter the climate became less arid and lakes appeared. Later a process of aridification started and an aeolian process finally replaced the marsh sedimentation. Owing to the insufficiency of dating data, identifying a reliable time scale from the profiles still remains a challenging task. But the glacial and periglacial sequences in the foot zone of the marginal highlands in the Alashan Highland provide a comparable chronology.

Roughly, the pediments dovetail the glacial-fluvial fans of the last glacial maximum (Hoevermann et al., 1998). This is a hint that the active formation of pediment took place during the last glacial period.

In general, if the old pediments and desert gorges in the region of present-day formation of the desert plains demonstrates higher precipitation in the past and a remnant landscape, then the aerodynamic relief indicating that the landscape patterns of desert plain and aeolian dunes are suitable to contemporary landforms corresponding to the present arid climate.

3.3. Palaeo-climate Change Deduced from Landscape Evolutions

The shift of a boundary between different landscapes and the transformation of landscape types are credible evidences of climate change, and the changing extents of different climatic controlled geomorphic landscapes help to provide estimates of the magnitude of climatic change (Hovermann, 1985; Gutierrez, 2005). Based on the modern climatic implication of the different landscape regions it is also possible to estimate past precipitation and temperature (Lehmkuhl and Haselein, 2000). For example, Hovermann et al. (1998) calculated the annual precipitation values of Qaidam region since 32 ka by comparing the annual sedimentation rate of clastic sediments in core CK 2022 from the central Qaidam Basin. Goudie (1983) points out that the peak of dust storm frequency occurs in areas where rainfall ranges between 100 and 200 mm/a. This is in accordance with many regional observations made in Tibet, Mongolia, and Africa (Hovermann, 1985, 1987; Hovermann et al., 1993; Lehmkuhl, 1997).

3.3.1. Aridity in the Basin

The transformations of old pediments and desert gorges to desert plains, the occurrence of wide shifting dunes and the existing large palaeo-lakebed areas all indicate the increasing aridity index in the Ejina Basin at present. In general, regional aridity is controlled by a conjunction of climatic, orographic and oceanographic factors. The zonal factor, continentality and the orographic effect are possible the main controls (Gutierrez, 2005).

According to the rule of climatology, the zonal factor occurs in relation to the trajectory of anticyclonic cells that control the tropical and subtropical deserts. Occasionally these deserts can be penetrated by storm systems, although precipitation is scarce. Zones of great aridity, surrounded by small semi-arid stripes, characterize them. The most representative example is the

Sahara desert. While the Ejina Basin is geographically located in the middle-latitude temperate zone of North Hemisphere, the zonal factor should not be the reason for its aridity. Continentality refers to the distance from ocean of the land. In temperate latitudes, the rain-producing fronts gradually lose their humidity along their way towards the interior of the continents. In these regions cold winters are common and the arid zones are bordered by semi-arid areas of great extension. The middle-latitude deserts of northern China and Central Asia have these characteristics. So it is an important factor controlling the aridity of the Ejina. The orographic effect occurs because of the presence of rain-shadow zones situated on the leeward side of mountainous chains, in regions with dominant winds, as in the Trade Winds belts or on some west coasts. When air masses descend the slopes of a chain of mountains they heat up and dry adiabatically. This orographic effect can be emphasized by the zonal situation and the continentality. Ejina Basin lies in the north edge of the NH Trade Winds belts. The Qilian mountain chains and the Tibetan Plateau on the southern part, together with the Alshan Highland and the Inner Mongolia Plateau on the eastern part of the Ejina Basin, act as such a giant rainfall barrier preventing the enter of vapor source from the Southeastern Asian Summer Monsoon and the Southwestern Asian Summer Monsoon.

We should also consider the albedo effect or reflectivity of the desert surfaces in the Ejina Basin because in some warm deserts of the world, the heating generates convection currents. In rocky and stone deserts (Gobi in the Ejina) the absorption of the warmth is maximum, it enhances the degree of evapotranspiration effect and induced the strength of aridity, especially if a cover of desert vanishes exists on it.

It is designated by geomorphologist (e.g., Meigs, 1953) that a desert is corresponds to a cold deserts when the mean temperature of the coolest month in it is less than 0 °C. The most significant characteristic of this environment is the great thermal annual oscillation that occurs. Accordingly, Ejina Basin is under a cold desert environment. In the desert studied, temperatures of the earth surface have been registered that vary between 25 °C and –15 °C (Figure 4 b). Precipitation is influenced by the topography and in the marginal highlands is frequently snowfall. Evaporation is low during the cold months and is commonly associated with strong winds. Due to its latitudinal position and to the anticyclonal domain, the Ejina deserts suffer a great insolation and the air temperatures can reach considerable values, as much as 40 °C. The warm winds in the desert regions produce the desiccation of many areas. These winds blow during the day and at night stay calm. They blow up from the desert to the more humid marginal zones and are commonly dust-laden.

3.3.2. Evidences from Pediments and Desert Gorges

Landscapes like pediment and desert gorge surrounding the Ejina Basin are also widely distributed around the eastern, southern and northern parts of the Alashan Plateau and in the desert areas of western China (Figures2-3). Relative to desert plain, these kinds of relict landforms in current hyper-arid and temperate climate context of western China are associated with more humid and colder climate in the past (Hovermann, 1985). According to studies of the sedimentary chronology in the central Alashan (Hoevermann et al., 1998; Yang, 2001), a humid and cold climate on the Alashan Plateau during the local glacial maximum of the last glaciation is consistent with the old formation of the pediments and desert gorges on the margin of this plateau. The thick gravel deposits of Quaternary Gobi deserts along the northern margins of the Tibet Plateau (Qilian Mountains) are middle and Upper Pleistocene ages (Liu et al., 1996). Field investigations have shown that there have been clear climatic fluctuations accompanied by landscape changes and paleohydrological variations in the desert areas of western China during the Late Quaternary (e.g. Chen and Bowler, 1986; Thompson et al., 1989; Fang, 1991; Li and Shi, 1992; Owen et al., 1997; Lehmkuhl and Haselein, 2000; Zhang et al., 2000; Ye and Ji, 2001; Yang et al., 2003, 2011a, b; Yang and Scuderi, 2010), and in these areas presently under formation of sand dunes and desert plains, underlying relict gorges and alluvial fans are attributed to the local glacial maximum (Yang, 2001; Yang et al., 2004). Based on the gemorphological and sediment properties of the regional comparability, especially the low degree of weathering (Zhu and Yang, 2009), it is reasonable to attribute the relic forms of desert gorges and pediments in the margin of the Ejina Basin to the local glacial maximum of the last glaciation (LGM). This opinion is also supported by the numerous data showing much wetter and colder climate during the Anaglacial (about 32-24 ka) in the Central Asia, northern China, Mongolia and on the Tibetan Plateau (Feng et al., 1998; Liu et al., 2002; Yang et al., 2004; Mischke et al., 2005; Herzschuh, 2006). For example, wide distribution of relict frost wedges and moraines in the Chaidamu and Tarim basins of China indicate that the coldest time of the last glaciation occurred at 30-25 ka in these two areas; the stratigraphy of dunes and high lake levels in the Badanjilin Desert and its vicinity regions in northern China and in Mongolia suggest a much more humid climate in western China and in Mongolia during marine oxygen isotope stage three (MIS 3) (Yang et al., 2004). Cold and wet climates were prevailed in the western part of the Chinese Loess Plateau during MIS 4 and MIS 3 (Feng et al., 1998). The northern boundary of the Gobi in Mongolia has shrunk owing to the wet climate as nine times during the past 40,000 yr, e.g. around 34,400, 30,700, 28,900, 24,500, 15,090, et al. (Feng, 2001).

Extensive pedimentation processes occurred during the last glaciation around the margins and central of the Alashan were identified (Hoevermann et al., 1998). Three different phases of pediment formation during the Pleistocene in the foreland of Qilan Mountains have also been distinguished (Hoevermann et al., 1998). For the periods of pediment development the annual precipitation in these areas should be considered to have been as high as 150-300 mm (Table 1). Presently the average annual precipitation varies from 110 mm in the southeast Alashan to 38 mm in the northwest and 36.5 in the Ejina Basin (Figure 4c) according to records obtained over the last five decades. Compared the current climate (40 mm in precipitation) with the corresponding climatic conditions of desert gorge (60-150 mm) and pediment (150-350 mm) formations, relicts of old pediments and desert gorges in the Ejina imply that the amplitude of precipitation variation in the past has reached 20 to 200 mm and this is much larger than present level. Giving the average annual precipitation was 60~350 mm on the margin of the Ejina Basin during the LGM, that would have been 50%~775% more than at present.

3.3.3. Evidences from Desert Plains (Gobi) and Alluvial Sedimentation

In arid regions Gobi deserts derived from alluvial fans are a common geomorphological feature (Goudie, 2009). The Gobi deserts (Figure 5) in the Ejina Basin are widely developed from alluvial sediments. Reconstructing the evolutionary history of Gobi desert has great significance in unraveling changes in both tectonic activity and climate (McFadden et al., 1987; Feng et al., 1998, 2001), because the formation and evolution of an alluvial fan are closely related to variations in the tectonic activity and climate. Abrupt floods and strong wind are two major dynamic factors in the formation of the Gobi Desert (Hovermann, 1985; Gutierrez, 2005). As a result, Gobi desert has experienced both hydraulic transportation and wind erosion and would record changes in tectonic, climate, and environmental conditions. Therefore, the beginning of the development of the Gobi Desert results in the end of alluvial processes.

Cosmogenic ^{10}Be in quartz gravel from the Gobi deserts of the Ejina Basin has been measured to assess the exposure ages of gravels (Zhang et al., 2006; Lv et al., 2010). The results showed that the Gobi desert in the northern margin of the Ejian basin formed 420 ka ago, whereas the Gobi desert that developed from alluvial plains in the central areas of the drainage basin came about during the last 190 ka (Figure 14), both of key periods during the late

Pleistocene glaciations. The latter developed gradually northward and eastward to modern terminal lakes of the river. These temporal and spatial variations in the Gobi deserts indicated a consequence of alluvial processes influenced by Tibetan Plateau uplift and tectonic activities within the Ejina Basin. Possible episodes of Gobi desert development within the last 420 ka indicate that the advance/retreat of alpine glaciers during glacial/interglacial cycles might have been the dominant factor to influencing the alluvial intensity and water volume in the basin. Intense floods and large water volumes would mainly occur during the short deglacial periods.

For intermontane basins considerable volumes of sediments have been stored as alluvial fan deposits or basin fills. Abundant sediments supply from Qilian Mountains by Heihe River results in the widespread distribution of alluvial fans along the mountain fronts and the Ejina basin. Both alluvial fans and alluvial cones consist mostly of gravel and sand but they can also contain boulders when debris flows contribute to their formation (Figure 6a-b). Alluvial deposition is tied to land surface instabilities caused by regional climate changes (Gutierrez, 2005). Alluvial fans in central Asia can be classified into two basic types: dissected and undissected; most dissected fans are composed of Pleistocene surfaces; whereas, most undissected fans are composed of late Pleistocene and Holocene surfaces (XETCAS, 1978). The distribution of fan surfaces formed in the mid-Pleistocene or earlier is much more limited (Oguchi et al., 2001). Well-preserved old surfaces are confined to areas where fan surfaces are being tilted rapidly toward fan apexes because of the relative uplift of fan toes. Alluvial fans in the Tarim Basin in southern Xinjiang, NW China are typical examples of this type (XETCAS, 1978). The alluvial fan in the Ejina basin is almost undissected. Compared the dissected and the undissected types of mountain-basin couples with the Tarim Basin, it possibly indicates that the alluvial sediments in the Ejina Basin are younger than those in the Tarim Basin, with alluvial sediments developed during late Pleistocene and Holocene.

3.3.4. Evidences from Palaeo-Riverbeds and Palaeo-lakebeds

Palaeo-landscape evolution and climate change can also be deduced from direct evidences of preserved river terraces and high lake levels in the Ejina Basin. Numerous existing relics of the palaeo-riverbeds and palaeo-lakebeds in the Ejina Basin still can be identified clearly at larger scale than present both from the morphology in the field and from the remote-sensing images (Figures2 and 13). It indicates that there have been large fluctuations in hydrological regime of the Ejina Basin during the part. According to the relic topography of the Ruoshui rive stem channel we can reasonably deduce that the terminal lake of the Ruoshui River should have been shifted from Gurinai to Juyanze, then to Sugunuoer and Gasunnuoer (Figure 2). Around these lake basins there are old lake terraces which are indicative of different lake levels. Taken lakebeds and lacustrine sedimentary profiles into consideration, it is possible to reconstruct lake level fluctuations from higher and lower lake levels in the Ejina Basin. Based on field investigation (Figure 6e) and geomorphological mapping of the Lake Sugunuoer depression, we calculated a high lake level of Sugunuoer at nearly 40 m above it present level. Along the southern shoreline of Guizihu three terraces are present. The highest terrace is ca. 40 m above the lake bottom. In the south of the Gurinai Lake basin the lake terrace lies 20 m above the bottom. Taking the bottom of Gushunnuoer as zero, around Sugunuoer and Juyanze the highest terraces are 34 m and 33.6 m above the bottom of Gushunnuoer respectively (Norin, 1980). As there are no basement mountains between Gushunnuoer and Sugunuoer, and there are lower topographic channels connecting Juyanze and Sugunuoer, it is reasonable to assume that these three lakes formed a unified single one during high level periods. The shells collected from the uppermost terrace of Sugunuoer were radiocarbon dated to 33,700±1300of the last glaciations (U-281; Norin, 1980).

Using radar wave data and Landsat images (TM, MSS), Guo et al., (2000) recognized a number of old river valley and lake basins buried by wind-blown sand in the north part of the Ejina Basin and the Alashan Highland. They delineated two parallel old drainage systems in the north and middle of the region and suggested that the moving sand belts mainly followed the old drainage courses. The old drainage verified that not only the Ejina Basin but also the Alashan Plateau was once a place occupied by many rivers and lakes in a warm and wet climate. Besides, the old drainage is NW-SE trending, not in accordance with the current south-higher-than-north and east-higher-than-west relief patterns. This fact indicates a relief reversion process caused possibly by neotectonic movement from Pleistocene epoch (Guo et al., 2000).

Palaeolake landscape evolution can also be in sighted from direct evidences of high lake levels in the Ejina Basin. Wunnemann et al. (1998) published results from sediments, lakeshore features and cores from Garsunnuoer and Sugunuoer lakes in the Ejina Basin, together with additional data from other locations. The chronology is based on 62 ^{14}C dates. Highest lake levels of +28 to +32 m occurred between 41 and ~33 ka in Garsunnuoer-Sugunuoer area. The lake level of Garsunnuoer was lowered 15 and 18 m at approximately 21 and 19 ka, respectively. Desiccation of Garsunnuoer-

Sugunuoer area took place after 19 ka with dune sand accumulated continuously until 14 ka. In addition, no datable material has been found between 18.6 and 12.8 ka. This supports hyperarid conditions, due to a lack of organic material. A freshwater lake was established at Garsunnuoer-Sugunuoer area after 11.3 ka.

The eastern Juyanze palaeolake in the east edge of the Ejina Basin (Figure 13d) (41.75-42°N, 101.5-102°E) provides evidence of a complex hydrological pattern during the Holocene (Hartmann and Wunnemann, 2009). Lithological, geochemical and mineral data of sedimentary cores from the lake indicate remarkable difference between the early Holocene and mid to late Holocene lake formation. At about 10,700 cal BP fresh water lake formation starts with extreme run off events under a wet climate. Extremely low run off occurred between 8900 and 8100 cal BP. Compared with the early Holocene wet climate, the mid to late Holocene local rainfall was less pronounced and displays a general decreasing trend towards complete desiccation of the lake (Hartmann and Wunnemann, 2009). A two-meter deep hole (39.83°N, 102.47°E, Alashan) above 15 m than Lake Huhejilin water surface indicate that the lake level of Huhejilin was at least 15 m higher from about 8,000 to 7,000 cal BP than at present (Yang, 2001).

The studies of lake evolution in the central, eastern Alashan Plateau and the vicinity basins (Mongolia) principally support the glacial palaeoclimatic interpretation of landscape evolution presented so far. Based on lake level studies, Pauchur et al. (1995) and Wunnemann et al. (2007a, b) presumed that rainfall in Inner Mongolia was considerably higher between 39 ka BP and 23 ka BP, and in the time around 13 ka BP than at present. Lacustrine carbonate found in the vicinity of Lake Cheligeri (39.88°N, 102.25°E, Alashan) was ^{14}C dated to 32,150±4920 cal BP (Yang, 2001). The occurrence of lake was considered to be synchronous with the Anaglacial, likely to reflect the earliest appearance of extensive lakes in the central Alashan. Wunnemann et al. (2007a, b) suggests that a superlative 50,000 km^2 freshwater lake existed on the northwestern margin of the Badanjilin Desert (Alashan) between 39 ka and 21 ka. During the period between 39 ka and 23 ka the climate in the Tenggeli Desert (Alshan) was also much more humid than at present (Pachur et al., 1995). About 33-23 ka BP there was a freshwater lake with an area of more than 16,200 km^2 and water depth of 25 m in the northwest of the present Tenggeli Desert (Zhang et al., 2002). The current Jilantai salt lake (39.74°N, 105.72°E, Alashan) was formed by the shrinkage of a very large lake (Zhang et al., 2002). Geomorphologic, sedimentological and geochronological data obtained from the Darhad Basin (50.3-52°N, 99-100°E, Mongolia) confirmed that the Darhad sedimentary sequence has been formed since the Pliocene and represents a detailed archive of environmental changes due to a high content of lacustrine beds (Krivonogov et al., 2012). There have been two stages of deep lake in this basin. The first lake was dammed by a glacier during late MIS 5 (Krivonogov et al., 2005) or, alternatively, during early to middle MIS 3 (Gillespie et al., 2008). The second damming, glacial or sedimentary, formed another deep lake during MIS 4 or MIS 2. The Holocene history of the Darhad paleolake was kept at low levels, during which the lake was minimal or disappeared at ca 12-9.6 and after 4.5 ka cal. BP, relatively deep at 9.6-7.1 and 6.4-4.5 ka cal. BP and shallow at 7.1-6.4 ka cal. BP.

3.4. Mechanism of Landscape Development and Climatic Fluctuation during the Last Glaciations

Figure 15 gives a summary of the different morphologenic processes and the deferent palaeoclimatic evidence in the basins and mountains of Central Asia during the last 100,000 years. This framework for climatic fluctuations is derived from references on palaeoclimatic events in Northern China and Mongolia (e.g., Lehmkuhl, 1997; Owen et al., 1997; Benn and Owen, 1998; Lehmkuhl and Haselein, 2000; Grunert et al., 2000; Yang et al., 2004, 2006). The glacier fluctuations are based on results from Tibetan glaciers (Benn and Owen, 1998; Lehmkuhl and Haselein, 2000; Liu et al., 2002). In Tibet and the surrounding mountains maximum glacier advances, which promoted the landscape development of desert gorges and pediment, took place in both stages of the Last Glaciation (MIS4 and MIS 2, e.g. 70-50 ka and 32-15/14 ka) (Benn and Owen, 1998) and for the Late Glacial around 15 ka. For the MIS 3 there were several dates on higher lake levels showing wetter climatic conditions. In the southern parts of the Mongolia Altai, two moraines of the last glacial cycle can also be found (Grunert et al., 2000). This landscape-climate model provides a key clue to understand the mechanism of landscapes development and climatic fluctuation in the Ejina Basin during the last glaciations.

The widespread distribution of Gobi desert of Central Asia originated mainly from alluvial fans and fanglomerates (pediment region). As shown in Figure 15 a transition from wetter (mainly MIS 3 and beginning MIS 2=Anaglacial period) to arid climatic conditions (OIS 2, Kataglacial period) occurred in most areas of Central Asia.

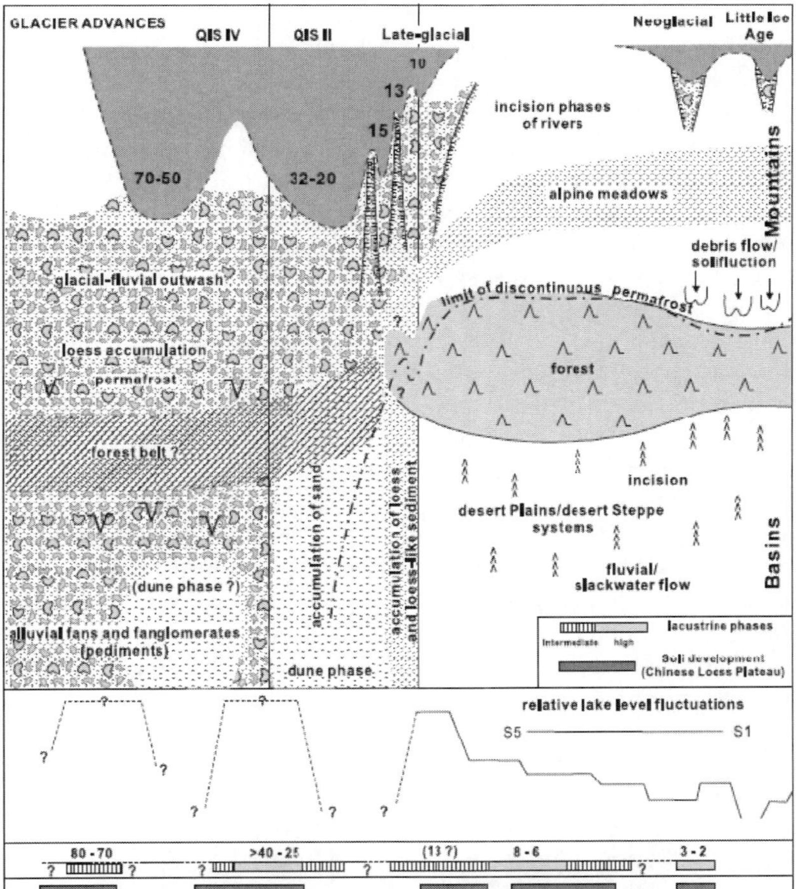

Figure 15. Landscape development including lake level changes, glacier fluctuations in the mountains (top) and basins (bottom) of northern China and Mongolia during the late Pleistocene (After Lehmkuhl and Haselein, 2000; Grunert et al., 2000).

Pollen and sedimentological evidence from western Loess Plateau indicates a smaller extent of the Gobi desert during cold and moist periods 73-60 ka and 27-19 ka and maximum expansion of the Gobi desert 60-50 ka and 19-10 ka (Feng et al., 1998). Observations on Late Quaternary alluvial fans in the Gobi of southern Mongolia (Owen et al., 1997) suggest that more humid conditions may have occurred between 40 and 23 ka followed by a period of increased aridity. Highest lake levels (28-32 m) occurred between 41 and ~33 ka in Garsunnuoer-Sugunuoer area (Norin, 1980; Wunnemann et al., 1998) of the Ejina Basin were related with higher levels of palaeoprecipitation and palaeohydrological cycle compared with present conditions.

In genesis, the increased precipitation in the western Inner Mongolia (Alashan and Ejina) during the last glaciation presumably originated from an increase of the westerlies (Yang et al., 2004). Benn and Owen (1998) also suggested that the glaciers of the northern Tibet and northern Karakorum responded to variations of the westerly winds during the Late Pleistocene. The Anaglacial time (ca. 32-24 ka) shows wetter conditions that could have been induced by an increase of the westerlies and a weakening of the winter monsoon (Lehmkuhl and Haselein, 2000). This first influenced the western part of Central Asia; the latter influenced the mainland of China and the eastern part of Central Asia. During Kataglacial times (ca. 24-15 ka) a strengthening of the winter monsoon due to a significant displacement of the Siberia high-pressure to the south could have caused the dry climatic conditions in north and central China (An et al., 1991; Ding et al., 1995; Pachur et al., 1995). The different variations, especially in precipitation, might have been dominated by the dynamic and thermal effects of the Siberia High. This could provide a significant increase in the gradient of the wind and temperature fields, with the high-pressure cell becoming more persistent in the winter and springtime. During more humid periods in Central Asia this pressure system might be relatively weak and farther north, enhancing cyclonic precipitation by the westerlies or/and the monsoon in springtime.

High lake level is evidence that it was considerably wetter in the central Alashan in the early Holocene (Yang 2001; Yang et al., 2004). However, Eastern Juyanhai (Sugunuoer) record of pollen data does not show the early Holocene humidity maximum (Herzschuh et al., 2004), as suggested by the sedimentary data from monsoon-influenced areas of China (Figure 1b). These results indicate an extremely variable moisture availability in arid regions of Inner Mongolia during the entire Holocene. Owing to the interplay of the East Asian Summer Monsoon (EASM) with the westerly waves at the boundary of the summer monsoon limit, variations from dry to wet climate conditions seem to have been more pronounced than in other monsoon regions of China. We can assume that efficient local rainfall and, thus, surface runoff are mainly linked to the interplay of the EASM and the westerly waves, as they are today (Figure 1b). An intensification of EASM may result in a seasonal northward shift of effective moisture (An et al., 2000; Wunnemann et al., 2007a, b), with more frequent contact with the westerly waves, thus producing local heavy rainfall in the desert regions with a significant variability at intra-seasonal, inter-annual and inter-decadal time scales (Ding and Chan, 2005). The wet climate during early Holocene in northern China is also explained by an asynchronous Holocene optimum (HO) in China (An et al., 2000), as the HO reaching a maximum at different times in different regions, e.g., ca. 10,000-8000 yr ago in northeastern China, 10,000-7000 yr ago in north-central and northern east-central China, ca. 7000-5000 yr ago in the middle and lower reaches of the Yangtze River, and ca. 3000 yr ago in southern China.

CONCLUSION

The formation of landscapes depends mainly on internal and external dynamics. The internal dynamics, especially the tectonic movements since Mesozoic and Cenozoic era, are responsible for the general framework of the present mountain-basin coupled landscape in Ejina. External dynamics are predominantly regulated by climate factors, which induced the geomorphological diversity of the landscape in the Ejina Basin. Desert plains, pediments, desert gorges, aeolian dunes, rive channels and lakebeds are the main existing geomorphological units in the Basin. Our study shows that there are strong links between climatic geomorphological processes and landforms and the complexity is further heightened by human activity. The landscapes and the boundary of different landscape types in the basin have undergone changes or transformation. In the sense of climatic geomorphology, the old pediment in the region of present-day formation of the desert plains demonstrates higher precipitation and frost weathering processes of wetter and colder climate in the past. Desert gorges surrounding the peripheral highlands of the basin are relic of glacial and periglacial processes dissected by glacier water. Floods and strong wind are two major dynamic factors in the formation of Gobi desert. Small dunes are distributed widely along the Ruoshui river banks. Typical aerodynamic relief often occurs in the area with an average annual precipitation of less than 30 mm. The average annual precipitation at present is ca 35 mm in the Ejina Basin, indicating that landscapes of desert plain and aeolian dunes are contemporary landforms corresponding to the present climate. Relicts of old pediments and desert gorges indicate that the amplitude of precipitation variation in the past has reached 20 to 200 mm and this is much larger than present level. These suggest that the basin-scale forms of existing landscape studied are the result of glacial, periglacial, fluvial and alluvial action, aggradation and degradational processes, as well as climatic changes. Previously studies have proved that the geomorphology of the central

Alashan Plateau is dominantly a consequence of Quaternary climatic change that has forced environmental changes in the mountains and forelands. Evidences from the Ejina Basin further confirmed it and the landscape evolutions in the Alashan Plateau have been influenced considerably by external dynamics (climate) since the Pleistocene. Studies about the palaeo-lake level changes on the lowest marginal areas of the basin show that the regional aridity has changed during the last glaciation period and the regional climate has experienced high precipitation during ca. 39-23 ka BP and great fluctuations during Holocene. The periods of lower aridity occurred at ca 33ka B.P., 23 ka B.P., 10 ka B.P. and during the early and middle Holocene. Due to the quite weak Asian summer monsoon during the glacial times, the periods of lower aridity during the late Pleistocene would arise from a probably stronger influence of westerlies on the arid areas of the northern China and Central Asia.

Acknowledgments

This work was financially supported by the Kezhen Young Talent Project of the IGSNRR-CAS (Grant no.: 2013RC101) and the National Natural Science Foundation of China (Grant nos.: 41371060 and 41271049). B.Q. Zhu is grateful to Prof. X.P. Yang for his illumination and inspiration on the author's the research work.

References

Ahnert, F., 1996. Introduction to Geomorphology. Arnold, London, p. 352.

An, Z., Wu, X., Lu, Y., Zhang, D., Sun, X., Dong, G., Wang, S., 1991. Paleoenvironmental changes of China during the last 18,000 years. In: Liu, T. (Ed.), *Quaternary Geology and Environment in China. Science Press,* Beijing, pp. 228-236.

An, Z., Porter, S.C., Kutzbach, J.E., Wu, X., Wang, S., Liu, X., Li, X., Zhou, W., 2000. Asynchronous Holocene optimum of the East Asian monsoon. *Quaternary Science Reviews* 19, 743-762.

Beggs, P.J., 2013. New directions: climatediversity: a new paradigm for climate science. *Atmospheric Environment* 68, 112-113.

Benn, D., Owen, L.,1998. The role of the Indian summer monsoon and the mid-latitude westerlies in Himalayan glaciation: review and speculative discussion. *Journal of Geological Society London* 155, 353-363.

Busche, D., 1998. Die Zentrale Sahara. *Oberflachenformen im Wandel.* Gotha, Perthes, p. 249.

Chen, K., Bowler, J., 1986. Late Pleistocene evolution of salt lakes in the Qaidam Basin, Qinhai Province China. *Palaeogeography, Palaeoclimatology, Palaeoecology* 54, 87-104.

Cooke, R.U., Warren, A., Goudie, A., 1993. Deserts Geomorphology. *UCL* Press, London, p. 526.

Ding, Y., Chan, J.C.L., 2005. The East Asian summer monsoon: an overview. *Meteorology and Atmospheric Physics* 89, 117-142.

Ding, Z., Liu, T., Rutter, N.W., Yu, Z., Guo, Z., Zhu, R., 1995. Ice volume forcing of Asian winter monsoon variations in the past 800,000 years. *Quaternary Research* 44, 149-159.

Fang, J.Q., 1991. Lake evolution during the past 30,000 years in China, and its implications for environmental change. *Quaternary Research* 36, 27-60.

Feng, Z., Chen, F., Tang, L., Kang, J.C., 1998. East Asian monsoon climates and Gobi dynamics in marine isotope stages 4 and 3. *Catena* 33, 29-46.

Feng, Z., 2001. Gobi dynamics in the Northern Mongolian Plateau during the past 20,000+ yr: preliminary results. *Quaternary International* 76/77, 77-83.

Fryberger, S., Dean, G., 1979. Dune forms and wind regime. In: McKee, E. (Ed.), A Study of Global Sand Seas. United States Geological Survey Professional Paper, 1052, Washington, pp. 137-169.

Gillespie, A.R., Burke, R.M., Komatsu, G., Bayasgalan, A., 2008. Late Pleistocene glaciers in Darhad Basin, northern Mongolia. *Quaternary Research* 69, 169-187.

Gordon, J.E., Barron, H.F., Hansom, J.D., Thomas, M.F., 2012. Engaging with geodiversitydwhy it matters. *Proceedings of the Geologists' Association* 123 (1), 1-6.

Goudie, A.S., 1983. Dust storms in space and time. *Progress in Physical Geography* 7, 503-530.

Goudie, A.S., 2009. Global deserts and their geomorphological diversity. In: Parsons, A.J., Abrahams, A.D. (Ed.), *Geomorphology of Desert Environments.* Springer Science + Business Media, pp. 9-20.

Grunert, J., Lehmkuhl, F., Walther, M., 2000. Palaeoclimatic evolution of the Uvs Nuur basin and adjacent areas (Western Mongolia). *Quaternary Internation* 65/66, 171-192.

Guo, H., Liu, H., Wang, X., Shao, Y., Sun, Y., 2000. Subsurface old drainage detection and paleoenvironment analysis using spaceborne radar images in Alxa Plateau. *Science in China Series* D 43, 439-448.

Guo, Z.T., Ruddiman, W.F., Hao, Q.Z., Wu, H.B., Qiao, Y.S., Zhu, R.X., Peng, S.Z., Wei, J.J., Yuan, B.Y., Liu, T.S., 2002. Onset of Asian desertification by 22 Myr ago inferred from loess deposits in China. *Nature* 416, 159-163.

Gutierrez, M., 2005. *Climatic Geomorphology.* Elsevier, Amsterdam.

Hartmann, K., Wunnemann, B., 2009. Hydrological changes and Holocene climate variations in NW China, inferred from lake sediments of Juyanze palaeolake by factor analyses. *Quaternary International* 194, 28-44.

Herzschuh, U., 2006. Palaeo-moisture evolution in monsoonal Central Asia during the last 50,000 years. *Quateranry Science Reviews* 25, 163-178.

Herzschuh, U., Tarasov, P., Wunnemann, B., Hartmann, K., 2004. Holocene vegetation and climate of the Alashan Plateau, NW China, reconstructed from pollen data. *Palaeogeography, Palaeoclimatology, Palaeoecology* 211, 1-17.

Hoevermann, J., 1985. Das System der klimatischen Geomorphologie auf landschaftskundlicher Grundlage. Zeitschrift fur Geomorphologie Neue Folge 56, *Supplementary Band,* 143-153.

Hoevermann, J., Hoevermann, E., Lehmkuhl, F., 1998. Geomorphologische Untersuchungen im noerdlichen Vorland des Qilian Shan, China. *Berliner geographische Abhandlungen* 63, 83-98.

Hovermann, J., 1987. Morphogenetic regions in Northeast Xizang (Tibet). In: Hovermann, J., Wang, W. (Eds), *Reports of the Qinghai-Xizang (Tibet) Plateau.* Science Press, Beijing, pp. 112-139.

Hovermann, J., Lehmkuhl, F., Portge, K.H., 1993. Pleistocene glaciations in eastern and central Tibet - Preliminary results of Chinese-German joint expeditions. *Zeitschrift fur Geomorphologie Neue Folge* 92, Supplementary Band, 85-96.

Krivonogov, S.K., Sheinkman, V.S., Mistryukov, A.A., 2005. Stages in the development of the Darhad dammed lake (Northern Mongolia) during the Late Pleistocene and Holocene. *Quaternary International* 136, 83-94.

Krivonogov, S.K., Yi, S., Kashiwaya, K., Kim, J.C., Narantsetseg, T., Oyunchimeg, T., Safonova, I.Y., Kazansky, A.Y., Sitnikova, T., Kim, J.Y., Hasebe, N., 2012. Solved and unsolved problems of sedimentation, glaciation and paleolakes of the Darhad Basin, Northern Mongolia. *Quaternary Science Reviews* 56, 142-163.

Lehmkuhl, F., Haselein, F., 2000. Quaternary paleoenvironmental change on the Tibetan Plateau and adjacent areas (Western China and Western Mongolia). *Quaternary International* 65/66, 121-145.

Lehmkuhl, F., 1997. The spatial distribution of loess and loess-like sediments in the mountain areas of Central and High Asia. Zeitschrift fur Geomorphologie Neue Folge 111, *Supplementary Band,* 97-116.

Li, Q., Wu, H., Yu, Y., Sun, A., Markovic, S.B., Guo, Z., 2014 Reconstructed moisture evolution of the deserts in northern China since the Last Glacial Maximum and its implications for the East Asian Summer Monsoon. *Global and Planetary Change 121*, 101-112.

Li, S., Shi, Y., 1992. Glacial and lake fluctuations in the area of West Kunlun mountains during the last 45,000 years. *Annals of Glaciology* 16, 79-84.

Liu, T., Ding, M., Derbyshire, E., 1996. Gravel deposits on the margins of the Qinghai-Xizang Plateau, and their environmental significance. *Palaeogeography, Palaeoclimatology, Palaeoecology 120*, 59-170.

Liu, T., Ding, Z., 1998. Chinese loess and the paleomonsoon. *Annual Review of Earth and Planetary Sciences* 26, 111-145.

Liu, T., Zhang, X., Xiong, S., Qin, X., Yang, X., 2002. Glacial environments on the Tibetan Plateau and global cooling. *Quaternary International* 97-98, 133-139.

Lu, H., Mason, J.A., Stevens, T., Zhou, Y., Yi, S., Miao, X., 2011. Response of surface processes to climatic change in the dunefields and Loess Plateau of North China during the late Quaternary. *Earth Surface Processes and Landforms* 36, 1590-1603.

Lv, Y., Gu, Z., Ala, A., Zhang, H., Goran, P., Lei, G., 2010. ^{10}Be in quartz gravel from the Gobi Desert and evolutionary history of alluvial sedimentation in the Ejina Basin, Inner Mongolia, China. *Chinese Science Bulletin* 55, 3802-3809.

Mabbutt, J.A., 1977. *Desert Landforms.* The MIT Press, Cambridge, MA, p. 340.

Mason, J.A., Lu, H., Zhou, X., Miao, X., Swinehart, J.B., Liu, Z., Goble, R.J., Yi, S., 2009. Dune mobility and aridity at the desert margin of northern China at a time of peak monsoon strength. *Geology* 37, 947-950.

Meigs, P., 1953. The world distribution of arid and semiarid homoclimates, *Riviews of Research on Arid Zone Hydrology*. UNESCO, Paris, pp. 203-209.

McFadden, L.D., Wells, S.G., Jercinovich, M.J., 1987. Influences of eolian and pedogenic processes on the origin and evolution of desert pavements. *Geology* 15, 504-508.

Mischke, S., Herzschuh, U., Zhang, C., Bloemendal, J., Riedel, F., 2005. A Late Quaternary lake record from the Qilian Mountains (NW China): lake level and salinity changes inferred from sediment properties and ostracod assemblages. *Global and Planetary Change* 46, 337-359.

Norin, E., 1980. Sven Hedin central Asia atlas. Memoir on maps, Stockholm, p. 114.

Oguchi, T., Saito, K., Kadomura, H., Grossman, M., 2001. Fluvial geomorphology and paleohydrology in Japan. *Geomorphology* 39, 3-19.

Owen, L.A., Windley, B.F., Cunningham, W.D., Badamgarov, G., Dorjnamjaa, D., 1997. Quaternary alluvial fans in the Gobi Desert, southern Mongolia: evidence for neotectonics and climate change. *Journal of Quaternary Science* 12, 239-252.

Owen, L.A., Richards, B., Rhodes, E.J., Cunningham, W.D., Windley, B.F., Badamgarov, J., Dorjnamjaa, D., 1998. Relic permafrost structures in the Gobi of Mongolia: age and significance. *Journal of Quaternary Science* 13, 539-547.

Pachur, H.J., Wunnemann, B., Zhang, H.C., 1995. Lake evolution in the Tengger Desert, northwestern China, during the last 40,000 years. *Quaternary Research* 44, 171-180.

Panizza, M., 2009. The geomorphodiversity of the Dolomites (Italy): a key of geoheritage assessment. *Geoheritage* 1, 33-42.

Peltier, L.C., 1950. The geographic cycle in periglacial regions as it is related to climatic geomorphology. *Annals of the Association of American Geographers* 40, 214-236.

Perucca, L., Martos, L.M., 2012. Geomorphology, tectonism and Quaternary landscape evolution of the central Andes of San Juan (30°S-69°W), Argentina. *Quaternary International* 253, 80-90.

Pope, G.A., Dorn, R.I., Dixon, J.C., 1995. A new conceptual model for understanding geographical variations in weathering. *Annals of the Association of American Geograpers* 85, 38-64.

Stevens, T., Lu, H., Thomas, D.S.G., Armitage, S.J., 2008. Optical dating of abrupt shifts in the Late Pleistocene East Asian monsoon. *Geology 36*, 415-418.

Sun, J., Ding, Z., Liu, T., 1998. Desert distributions during the glacial maximum and climatic optimum: example of China. *Episodes 21(1)*, 28-31.

Sun, J., 2002. Provenance of loess material and formation of loess deposits on the Chinese Loess Plateau. *Earth and Planetary Science Letters* 203, 845-859.

Testa, B., Aldighieri, B., Bertini, A., Blendinger, W., Caielli, G., de Franco, R., Giordano, D., Kustatscher, E., 2013. Geomorphodiversity of the San Lucano Valley (Belluno Dolomites, Italy): a well-preserved heritage. Geoheritage, 5, 151-172.

Thompson, L.G., Mosley-Thompson, E., Davis, M.E., Bolzan, J.F., Dai, J., Yao, T., Gundestrup, N., Wu, X., Klein, L., Xie, Z., 1989. Holocene-Late Pleistocene climatic ice core records from the Qinghai-Tibetan Plateau. *Science* 246, 474-477.

Wunnemann, B., Pachur, H.J., Li, J., Zhang, H., 1998. Chronologie der pleistozanen und holozanen Seespiegelschwankungen des Gaxun Nur/Sogo Nur und Baijian Hu, Innere Mongolei, NW-China. *Petermanns Geographische Mitteilungen* 142, 191-206.

Wunnemann, B., Hartmann, K., Altmann, N., Hambach, U., Pachur, H.J., Zhang, H., 2007. Interglacial and Glacial fingerprints from lake deposits in the Gobi Desert, NW China. In: Sirocko, F., Claussen, M., Sanchez-Goni, M.F., Litt, T. (Ed.), *The Climates of the Past Interglacials. Developments in Quaternary Sciences,* vol 7. Elsevier, Amsterdam, pp. 323-347.

Wunnemann, B., Hartmann, K., Janssen, M., Zhang, H.C., 2007. Responses of Chinese desert lakes to climate instability during the past 45,000 years. In: Madsen, D.B., Chen, F.H., Gao, X. (Ed.), Late Quaternary Climate Change and Human Adaptation in Arid China. *Developments in Quaternary Science,* vol 9. Elsevier, Amsterdam, pp. 11-24.

XETCAS (Xinjiang Expedition Team of the Chinese Academy of Sciences), IGCAS (Institute of Geography of the Chinese Academy of Sciences), DGBNU (Department of Geography of Beijing Normal University), 1978. *Geomorphology in Xinjiang.* Science Press, Beijing (in Chinese).

Yang, S., Ding, Z., 2008. Advance-retreat history of the East-Asian summer monsoon rainfall belt over northern China during the last two glacialeinterglacial cycles. *Earth and Planetary Science Letters* 274, 499-510.

Yang, X., 2001. Late Quaternary evolution and paleoclimates, western Alashan Plateau, Inner Mongolia, China. *Zeitschrift fur Geomorphologie Neue Folge* 45, 1-16.

Yang, X., Liu, T., Xiao, H., 2003. Evolution of megadunes and lakes in the Badain Jaran Desert, Inner Mongolia, China during the last 31,000 years. *Quaternary International* 104, 99-112.

Yang, X., Rost, K.T., Lehmkuhl, F., Zhu, Z., Dodson, J., 2004. The evolution of dry lands in northern China and in the Republic of Mongolia since the Last Glacial Maximum. *Quaternary International* 118-119, 69-85.

Yang, X., Preusser, F., Radtke, U., 2006. Late Quaternary environmental changes in the Taklamakan Desert, western China, inferred from OSL-dated lacustrine and aeolian deposits. *Quaternary Science Reviews* 25, 923-932.

Yang, X., Scuderi, L.A., 2010. Hydrological and climatic changes in deserts of China since the late Pleistocene. *Quaternary Research* 73, 1-9.

Yang, X., Scuderi, L., Liu, T., Paillou, P., Li, H., Dong, J., Zhu, B., Jiang, W., Jochems, A., Weissmann, G., 2011a. Formation of the highest sand dunes on Earth. *Geomorphology* 135, 108-116.

Yang, X., Scuderi, L.A., Paillou, P., Liu, Z., Li, H., Ren, X., 2011b. Quaternary environmental changes in the drylands of China: a critical review. *Quaternary Science Reviews* 30, 3219-3233.

Yang, X., Li, H., Conacher, A., 2012. Large-scale controls on the development of sand seas in northern China. *Quaternary International* 250, 74e83.

Ye, W., Ji, F., 2001. Comparision of paleoclimatic characteristics between monsoon and westerly areas, China. *Chinese science bulletin* 46 (Supplementary), 119-124.

Zhang, H., Ma, Y., Peng, J., Li, J., Cao, J., Qi, Y., Chen, G., Fang, H., Mu, D., Pachur, H.J., Wunnemann, B., Feng, Z., 2002. Palaeolake and palaeoenvironment between 42 and 18 ka BP in Tengger Desert, NW China. *Chinese Science Bulletin* 47, 1946-1956.

Zhang, H., Ma, Y., Wunnemann, B., Pachur, H., 2000. A Holocene climatic record from arid northwestern China. *Palaeogeography, Palaeoclimatology, Palaeoecology* 162, 389-401.

Zhang, H., Ming, Q., Lei, G., Zhang, W., Fan, H., Chang, F., Wunnemann, B., Rtmann, K., 2006. Dilemma of dating on lacustrine deposits in a hyperarid inland basin of NW China. *Radiacarbon* 48, 219-226.

Zhu, B., Yang, X., 2009. Chemical Weathering of Detrital Sediments in the Taklamakan Desert, Northwestern China. *Geographical Research* 47, 57-70.

Zhu, B., Yu, J., 2014. Aeolian sorting processes in the Ejina desert basin (China) and their response to depositional environment. *Aeolian Research 12*, 111-120.

Zhu, B., Yu, J., Rioual, P., Ren, X., 2014a. Particle size variation of aeolian dune deposits in the lower reaches of the Heihe River basin, China. *Sedimentary Geology 301*, 54-69.

Zhu, B., Yu, J., Rioual, P., Gao Y., Zhang, Y., Min, L., 2014b. Geomorphoclimatic characteristics and landform information in the Ejina Basin, Northwest China. *Environmental Earth Sciences*, doi: 10.1007/s12665-014-3927-9.

In: Arid and Semi-Arid Environments
Editor: Gabriel Ferguson
ISBN: 978-1-63483-383-7

Chapter 4

New Utilization/Conservation Dilemmas in the Karoo, South Africa: Potential Economic, Demographic and Entrepreneurial Consequences

D. F. Toerien[*]
Centre for Environmental Management, University of the Free State, Bloemfontein, South Africa

Abstract

The landscape of the Karoo, the arid to semi-arid heartland of South Africa, has for millennia been subjected to a conservation/utilization dilemma. At first hunter-gatherer San tribes lived off Karoo plants and animals. Some two thousand years ago Khoikhoi herders and their livestock arrived. In the 18th Century European farmer settlers moved into the Karoo and replaced the San and Khoikhoi. In time they became wool producers and suppliers to Industrial Britain.

Wool was a major export staple of the Cape Colony. After World War II wool passed its export pinnacle and by the 1960s some Karoo towns were regressing. Increased poverty was a big concern and numerous government-instigated investigations followed. Since then some Karoo towns have prospered in post-productive fashion.

[*] D. F. Toerien. Centre for Environmental Management, University of the Free State, Bloemfontein, South Africa 9300. E-mail: Toeriend@ufs.ac.za.

Now the Karoo faces new development challenges. South Africa's part of the Square Kilometre Array, a large radio telescope, is being developed in the western Great Karoo. Some large solar power installations have also been erected. The Karoo was also identified as a likely source of shale gas. The South African government recently decided to allow exploration to proceed, an action that has occasioned vociferous opposition and rekindled the conservation/utilization dilemma. Exploration could last a decade and the period should be used to gather facts needed for wise and transparent decisions. This chapter contributes to this by focusing on the links between economic value addition, demographics, personal income and entrepreneurship in selected Karoo towns. Statistical and other techniques are used to quantify relationships between gross economic value addition, town populations, total personal income and total number of enterprises in Karoo towns. A power law describes the enterprise richness of Karoo towns and an analysis of functional business diversity indicates a tipping point at about 100 enterprises. The ways by which entrepreneurial spaces are used in Karoo towns are elucidated and a predictive capability is developed about the likely responses of populations, entrepreneurs and employment in Karoo towns to economic impacts, be they from the exploitation of shale gas or other developments. A lack of local information forced the use of US information about drilling rig numbers and densities in shale gas and oil extraction. This enables predictions indicating substantial economic and other benefits to some Karoo towns from shale gas extraction. These benefits have to be weighed against potential detrimental impacts of gas extraction in order to seek possible win-win solutions.

INTRODUCTION

South Africa's large semi-arid to arid central heartland, a high plateau that is generally known as the Karoo, ranges some 600 km from east to west and some 600 km from north to south [1] and covers about one-third of the country.

Man has impacted upon the Karoo for a long time and different phases of human occupation have influenced the Karoo biome in different ways. At present a number of new impacts have appeared or are potentially on the way, in particular: shale gas exploitation, the Square Kilometre Array (SKA), a large international scientific project, and projects for the generation of solar power. Shale gas exploitation, should it proceed, could impact a very large area of the Karoo. It is, therefore, necessary to consider to what extent these new developments could impact on Karoo towns and their residents.

Towards the end of 2014 there was a precipitous drop in the world oil price, which was at least partly ascribed to the influence of the shale revolution in the United States where in the space of six years oil production has nearly doubled to nine million barrels a day from five million a day [2]. Oil and gas production from shales has, therefore, become strategically important. South Africa has one major sedimentary basin, the Karoo Basin, which contains thick, organic-rich shales with gas potential [3]. The ancient Karoo Basin is large (~ 600000 km^2, see Figure 1) extending across nearly two-thirds of South Africa, with the southern portion of the basin, which is potentially favourable for shale gas, positioned in the area now known as the Karoo.

As elsewhere in the world, potential gas extraction from Karoo shales has elicited vehement debate [4]. De Wit [4] points out that two camps have emerged. On one side there are those who regard Karoo shale as delivering new solutions to meet intergenerational equity and who favour the extraction of shale gas should it be economically feasible.

On the other hand conservationists argue that the extraction of gas will leave massive irreparable environmental scars on one of South Africa's iconic landscapes and some of them have pointed to possible human health hazards from associated chemical pollution.

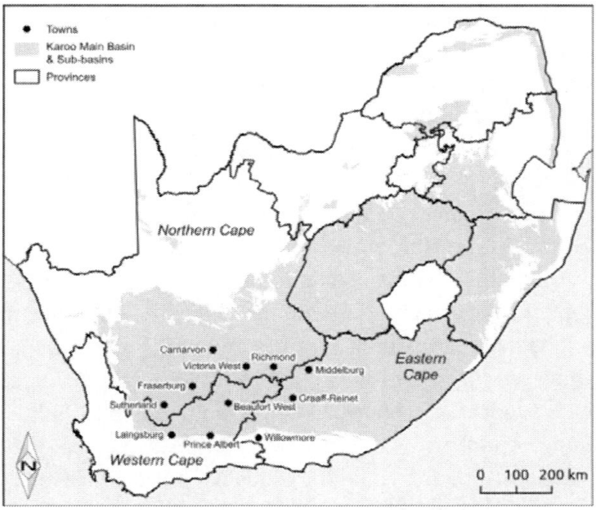

Figure 1. The sedimentary Karoo Basin. The names of the towns indicate the part of the Karoo where shale gas might be present.

According to this camp, South Africa would be better off leaving the gas in the ground.

In 2011 the South African Department of Mineral Resources halted new applications for exploration rights for shale gas in order to investigate the impact that shale gas exploration could have on the environment. An interdepartmental task team was set up to head this process [5].

At the time of writing this chapter the department was still working on a set of regulations that would ensure that shale gas exploration did not threaten South Africa's environment or compromise research projects linked to the SKA.

The Karoo is a place of unique biodiversity, stark beauty, wide open vistas and unsurpassed night skies [4]. De Wit [4] states that no one should want to set out to ruin that. However, the Karoo is also a place of intense poverty, with marginalized structurally unemployed people and some of the greatest chasms between the rich landowners and those who own nothing. For instance, De Jongh [6] discusses the plight of the 'Karretjie People' who by the turn of the previous century were confined to their temporary shelters on the verges of country roads. They had no land, or even free access to any space or place, and had remained at best largely socio-economically 'invisible' to the local population or, at worst, strangers in their own country. The 'natives of the Karoo' have before been challenged or their landscape altered by commerce without their consent [4].

A dilemma is a pair of apparently contradictory statements [7: 16]. In Greek *dilemma* means literally 'two propositions'. In essence, the debate about the exploitation of shale gas in the Karoo is a classic example of the dilemma between the utilization of a resource versus its conservation. Construction of a very large scientific project such as the SKA and of large installations to capture solar power in the Karoo also utilize the landscape of the Karoo, and also form part of the utilization/conservation dilemma of the Karoo.

Oil companies apparently fail to understand why the public and environmental lobbies mistrust their motives and plans to exploit shale gas, no matter what they say or promise [4]. De Wit [4] also indicates that it is necessary to understand the motives of large landowners to halt exploitation. These issues suggest that there is little progress towards synergetic solutions and that there is a lack of information, probably on many levels: ecological, socio-economic, cultural, etc. [8]. In 2014 the then deputy director general of the South African Department of Mineral Affairs, Mosa Mabuza, briefed a parliamentary committee on shale gas exploration licenses [9]. He indicated that hydraulic fracturing (fracking) could start 12 to 18 months after licenses were issued.

The construction of MeerKAT, an initial phase of the SKA radio astronomy project, which will consist of 64 large dish antennas, has already commenced. It is expected that 16 antennas will be in place by the end of 2015 and be ready to begin research. By the end of 2016 all 64 antennas will be in place [10]. Brits [11] explains that in time the SKA will be expanded to include 256 antennas and eventually 2500, many of which will be in countries other than South Africa (e.g. Australia, Ghana, Botswana, Kenya, Madagascar, Mauritius, Zambia, Mozambique and Namibia). Each of these antennas will be four storeys high and require precision building to within 1 mm tolerances. The project will require connections to extremely powerful computers. Clearly, high-level manpower will also be needed locally during the building and operation of the radio telescopes.

A number of large solar power installations have also been constructed in the Northern Cape and additional ones are planned [12]. During construction and operation these are also locally dependent on skilled workers.

All of these developments will impact (or are already impacting) the Karoo landscape and its residents. The question is how and to what extent? The window of opportunity to gather more information to create synergy before political and business decisions are made, in particular about shale gas exploitation, is short and should be utilized effectively.

One area that appears to be scientifically generally underexplored is how such developments will impact the socio-economic situation of towns. Even in the US, the major exploiter of shale gas in the world, what happens to the economy of towns when shale oil and gas exploitation takes place in their vicinities is not well quantified and only the broad economic impacts of shale oil and gas exploitation on regional economies have been analyzed. Deutch [13] and Morse [14] comment in general on the good news resulting from the exploitation of shale gas and oil. IHS Global Insight [15] prepared a report for the American Petroleum Institute on the economic and energy impacts of proposals to regulate hydraulic fracturing in the US. The economic impacts of fracking in Pennsylvania have been studied [16; 17] but Kinnaman [18] evaluated and criticized aspects of these studies. Hammond [19] studied economic impacts in West Virginia and Weber [20] studied some counties in Colorado, Texas and Wyoming.

Anecdotal information over a period in a newspaper, the New York Times, describes some impacts at town level. On the positive side, Krauss [21] remarks that a 17-mile stretch of road between a forsaken South Texas village,

Catarina, and the county seat of Carrizo Springs, was until 2010 a patchwork of derelict gasoline stations and rusting warehouses. By 2011 the region was in the hottest new oil play in the country, with giant oil terminals and sprawling RV parks replacing fields of mesquite. Each rig involved in exploitation in Texas represented about 100 jobs, from roughneck field hands to maintenance workers. Restaurants were full and the traffic was heavy around the town of Midland. Cutting back on rigs would negatively impact on the economy of the town [21].

In Western Pennsylvania the fortunes to be made salved the headaches and heartaches about shale exploitation in a stretch of Gasland [22]. Weisman [22] remarks that whole towns had been woken from the dead. However, he also tells the story of a family where the wells on their property were spinning off an estimated $20,000 a month in royalties, not for them, but for the New Jersey person who sold them the property but kept the gas rights.

An oil and gas stampede was transforming towns from the green hills of western Pennsylvania to the plains of North Dakota and eastern Colorado, bringing a flood of money, jobs and attendant environmental concerns [23].

However, on the negative side, in western Colorado, a region rich in natural resources and where oil and gas jobs formed the bedrock of the local economy, the boom had dried up. Main Street businesses were struggling and big new schools built to accommodate a surge of students from the previous energy rush were watching their enrolments dwindle. The cause of their woes was a tumble in gas prices due to increased competition [23].

A turnaround in the Rust Belt of Ohio was part of a transformation spreading across the heartland of the US driven by a surge in domestic oil and gas production that changed the economic calculus for old industries and downtrodden cities alike [24]. Entire sectors, such as manufacturing, hotels, real estate and even law, were being reshaped.

These newspaper anecdotes clearly show that on one hand towns have benefited from gas and oil extraction from shales in the US. However, negative impacts were also recorded. Neither the beneficial nor the detrimental impacts were precisely quantified. One can, therefore, broadly postulate that shale gas exploitation, the SKA project and the construction of large-scale solar power installations will probably benefit Karoo towns economically. However, apart from potential negative environmental impacts, cessation of gas extraction will diminish the flow of money into the local economies and thus negatively influence the towns.

The creation of synergetic solutions to deal with the utilization/conservation dilemma will need more quantified information than just broad statements. Daniel Kahneman, a Nobel economics laureate, discusses and praises the predictive powers of simple algorithms, which are often more accurate than the predictions of experts [25: 226].

Economic, demographic and enterprise regularities in the form of highly significant correlations have been observed in Karoo towns as well as many other South African towns [e.g. 26, 27, 28]. The observed regularities can be used to predict the interlinked economic, demographic and entrepreneurial dynamics in South African towns [29], a possibility acted upon in this contribution.

The primary goals of this contribution, therefore, are: (i) to determine if economic, demographic and/or entrepreneurial proportionalities (regularities) are present in Karoo towns that might be affected by the new developments, and (ii) should proportionalities be present, to use them to predict the extent of potential future development impacts on the local economies of affected Karoo towns. Special attention is given to possible economic impacts of fracking of Karoo shales for the extraction of gas.

This chapter has three parts and the golden thread is the link between man's historic impacts on the Karoo (part 1) with various economic, demographic and entrepreneurial proportionalities (regularities) observed in South African towns (part 2) in order to develop some understanding of the potential impacts of new developments in the Karoo (part 3).

Part 1 provides context and it starts with a short review of the Karoo biome, followed by a discussion of the historic impacts of man on the Karoo, tracing the history in the Karoo of hunter-gatherers, Khoikhoi herders, European traders and subsistence colonist farmers (also called trekboers), who later morph into commercial wool farmers. Thereafter the initial urbanization of the Karoo and eventual concerns about the regression of Karoo towns are presented and the socio-economic realities are discussed.

In part 2, enterprise development in Karoo towns is reviewed and proportionalities in the enterprise structures of Karoo towns and their implications are considered. This involves a re-examination of the relationships between economic value addition, population dynamics and enterprise development in Karoo towns. Further statistical and other analyses of the relationships between the economic, demographic and enterprise structures of Karoo towns are described.

In part 3, information about the extraction of shale gas and oil in the US is combined with quantified relationships between economic value addition, population sizes and enterprise structures of Karoo towns in order to make predictions of the future impact of shale gas extraction (if it is approved) on Karoo towns.

A conclusion concludes the chapter.

PART 1

The Karoo Biome

The geology and physiography of the Karoo vary a lot. There is a western escarpment that incorporates the Roggeveld, Nieuveld and Sneeuberg mountains, which may exceed 2500 m in height. Inland to the east lies the vast frosty peneplain of the Upper Karoo [30]. The climate of the Karoo Biome is arid to semi-arid, separated into different rainfall areas. The western part receives more than 60 percent of its rain from cyclonic fronts in winter. The even rainfall area receives rain any time of the year, while the large north eastern part receives summer rainfall, mostly by way of thunderstorms. Rainfall reliability is greatest in the southern and eastern parts and lowest in the western and northern parts. Mean rainfall varies from 100 mm in the west to more than 400 mm in the east, but for most of the biome it rarely exceeds 250 mm. Snow in winter is often recorded on the high areas of the escarpment. Mean annual temperatures vary from about 22.5°C in the north-west to 15°C on the escarpment. The Karoo vegetation is characterized as dwarf open shrubland. In the Great Karoo succulents are less conspicuous and dwarf deciduous shrubs predominate. Grasses may be common and even dominate, depending on soil, rainfall and grazing intensity [30].

Man and the Karoo

For many millennia the Karoo was utilized by hunter-gatherers, the/Xam-speaking San who moved around according to the availability of natural resources. Two to three thousand years ago Khoikhoi herders [31: 2, 32: 4], also known as Khoekhoen [6], moved in from the north (the name Khoikhoi will be used in this chapter).

The Khoikhoi were pure pastoralists and planted no crops other than dagga (marijuana) [32: 5]. They kept straight-backed and long-horned cattle, fat-tailed woolless sheep and goats [33: 48]. They had to move constantly and disperse widely in search of fresh pastures and their society was not based on the possession of land. The relationship between the herders and the hunter-gatherers was not always peaceful because of the San's propensity to steal animals from Khoikhoi herds [31: 27] and the Khoikhoi herds' dependency on the resources used by the game that the San hunted and depended on.

Due to higher rainfall, the more luxuriant pastures of the south-western coastal area of the African continent could support immense Khoikhoi herds and flocks. These pastures and the animals on them were to make the Khoikhoi objects of envy when a new force, Europeans, arrived from overseas [32: 7].

The European thrust into the lands of the Khoikhoi consisted of three distinct, though overlapping, frontiers that were initiated respectively by: (i) traders, (ii) cultivators and (iii) the trekboers [32: 7]. Elphick uses the term frontier to indicate a region where regular contact takes place between two or more culturally distinct communities, and where at least one of the communities is attempting to control the other but has not yet completely succeeded in doing so.

The first livestock trade between Khoikhoi and Europeans probably happened on Saturday, 2 December 1497 [31: 55]. By the 1590s ships began to put in regularly at Table Bay en route between Europe and the Indies and the trading frontier was on its way [32: 9]. The Khoikhoi were willing to supply large quantities of beef and mutton through bartering. The European agents were employees of trading companies of the various trading nations. In this way the Khoikhoi became linked into a global trading system. From 1652, when the Dutch East India Company (Vereenigde Oost-Indische Compagnie or VOC) founded a permanent settlement at Table Bay until about 1700, the trading frontier expanded from Table Bay throughout the south-western Cape. In the 18th century it moved hesitantly further into the interior [32], and also into the Karoo. A system of bartering at Table Bay was replaced by bartering expeditions to inland Khoikhoi tribes [32].

The inability and/or unwillingness of the Khoikhoi to supply enough livestock, particularly cattle, to supply the ships and the settlement at Table Bay, led to a frontier of agrarian settlement [32: 11]. The VOC released some of its employees from their contracts and set them up as independent farmers. These so-called free burghers were to sell grain and meat to the VOC at fixed prices. Friction between the Khoikhoi and the free burghers was almost inevitable and two Dutch-Khoikhoi wars followed, which increasingly led to the subjugation of the Khoikhoi tribes. The expansion of the colony into outlying districts intensified the need of Khoikhoi labour on the farms. By the mid-1690s some Khoikhoi were in permanent employment and lived with their families and livestock on European farms [32: 17].

The trekboer frontier started ca.1700 following the settlement of the Tulbagh valley. This was the first region of settlement whose climate, terrain, and distance from markets favoured a chiefly pastoral rather than a mixed agricultural economy [32: 21]. As the number of stock farmers increased more and more loan farms were taken out at ever increasing distances from Cape Town [34: 58]. The direction of trekboer expansion was largely a function of the nature of the terrain, the availability of permanent surface water, and the quality of pasture [34: 58-59]. From the 1750s to 1770s many areas of the Karoo were occupied by trekboers at the expense of the Khoikhoi [34: 59]. By 1780, the trekboers had taken possession of virtually the entire Karoo [35: 6]. Guelke [34: 59] states that: "The country almost certainly supported more nomadic Khoikhoi per km² in the seventeenth century than trekboers in the eighteenth."

The lives of the trekboers in the Karoo under the VOC were extremely hard [34: 62). The poorly developed transportation system and low population densities combined to prevent significant labour specialization and urban development. In the absence of artisans and other specialists, the inhabitants forfeited many of the economic advantages and social amenities derived from a division of labour. For the most part they did without physicians, blacksmiths, wagon makers, masons, teachers, etc. and had to be jacks-of-all-trades. The gradual disintegration of the Khoikhoi communities in the interior, however, provided the trekboers with cheap labour. The system of loan farms for which a small annual rent was paid promoted the over-use of the natural resource base of the region. Guelke [34: 66] shows that whereas a pastoral estate was the equivalent of 40.4 percent of an arable estate in the 1730s, it had reduced to only 11.7 percent by the 1770s.

Guelke [34: 70] remarks:

> [F]arm improvements waited on better transportation; improvements in transportation waited on an increase in population density; an increase in population density waited on farm improvements. As land was abundant in the open frontier, here was a vicious circle from which there was no escape. Lacking the basic requirements for sustained economic growth, the Cape interior remained economically stagnant for the greater part of the eighteenth century.

In 1795 the British took temporary control of the Cape. This was repeated in 1806 and control was not again relinquished. In 1814 the Cape formally became a British colony [36: 86]. Tamarkin [37: 15] remarks: "The absorption of the Cape by the foremost modern industrial state, which was also the dominant imperial power, set in motion processes of change which transformed the scale, the scope, and the nature of the economic order, and the physical and demographic landscape of the country."

Agriculture continued to be the foundation of the Cape economy and it was dominated by a farming community whose origin was in the VOC period. Under the British the Cape had to find and exploit profitable niches if it were to benefit from its British connection, and the production of wine, grain and livestock for foreign markets, especially the growing British market, had to be its economic mainstays [37: 17].

From the 1840s the colonial government improved the road infrastructure of the Cape and linked the Cape more firmly to the world market, launching a course of progress and relative prosperity [37: 19]. The export of wool transformed the Cape economy and was driven by the growing demands of the wool industry in Britain. Wool production in the Karoo, which previously had been relatively marginal, expanded. The subsistence trekboers of the Karoo slowly converted to wool farming. Tamarkin [37: 18] remarks:

> There developed a commercial chain leading from London through major urban centres, to regional centres and then to local dorps [towns]. At the end of this chain was the smous [pedlar], the itinerant trader, who brought the message of the market economy to the most isolated farmer. The story of the Mosenthal brothers who, after having established themselves, opened a branch in Graaff Reinet which in turn served as a centre for sub-branches in outlying centres like Richmond, Murraysburg, Burgersdorp, Hopetown and Aliwal North, was by no means unique.

This reference to Karoo towns showed that these towns formed part of the chain. The development of the wool industry ushered in the first real economic boom in the Cape.

By 1865 there were eight million wool sheep in the Cape and by 1891 13.5 million. The Anglo-Boer War (1898 – 1902) severely damaged the wool industry in South Africa but by 1911 the number of sheep had more than recovered [33: 67]. At this time the limits of the capacity of available pasture to support sheep had been reached and overgrazing and trampling were not only aggravating the danger of erosion and the effects of drought, but also began to reduce the carrying capacity of pastures [33: 67]. Nevertheless, by 1925 South Africa was the fourth largest wool producer in the world and after

World War II higher prices encouraged further expansion until the mid-1960s [33: 67]. Overstocking and overgrazing had, however, become huge problems.

A Commission of Enquiry into Agriculture stated in 1968: "The consequences of injudicious utilization of the natural vegetation and inefficient grazing management have been that large areas of the desert veld and arid Karoo have turned into desert and that dry sweet-veld areas have changed into Karoo and macchia..." [33: 47-48]. Utilization outstripped conservation. By the middle of the 20th Century there was a trend towards fewer farms in the Karoo. Between 1936 and 1960, the farming population in the central Karoo virtually halved, from 15,925 to 8,110 [35: 17]. The Karoo which once was the economic heartland of South Africa by virtue of its extensive rangeland agriculture that dominated the rural-based economy of the country in the 19th century, had by the 20th century come to be perceived as being a dry wasteland of little value, in which farming practices were often the direct or indirect cause of environmental degradation [38].

More recently farming in the Karoo has come to be dominated by a reduced number of skilled and well-resourced farmers.

Man's impact on the Karoo had progressed from hunter-gatherers who lived off game and plants, to Khoikhoi herders and trekboer subsistence farmers who lived off their livestock, to commercial wool farmers who delivered an export staple, but who eventually overstocked and overgrazed the natural ecosystem of the Karoo.

The challenges of the utilization/conservation dilemma were ignored and the basic ecosystem functions severely impaired. However, this was not the last of man's impacts, although the focus had shifted.

Urbanization of the Karoo

Under the VOC the Cape long remained agrarian in nature and the scarcity of towns reflected this [39: 25]. Besides Cape Town there were only eight other towns by the end of the VOC rule in 1806 [calculated from 39: 361] and only one, Graaff Reinet, was in the Karoo. Founded in 1786 for administrative purposes, Graaff Reinet was not much of an urban centre even by the early 1800s [37: 16]. The Karoo was only important to the VOC as a source of livestock to supply its fleets and the community at Cape Town, not as part of a colony that had to be managed in the long term.

Following the British occupation of the Cape and the subsequent wool export-driven economic boom in the mid 19th century, the colonial government promoted the rapid growth of villages and towns across the Cape [37: 16, 23]. These towns were founded mostly for religious and/or administrative purposes [39: 25]. This was also the case in the Karoo [40: 26]. Fransen [39: 361-362] lists at least 35 towns that were founded in the Karoo between 1806 and 1900. Small Karoo towns served as service centres to the surrounding farming communities and the population remained socially and economically isolated [35: 7; 41, 185]. Rural hinterlands and their respective towns had a mutually dependent relationship, which lasted for a long time. However, by the 1970s there were serious concerns regarding the regression of many small Karoo towns [e.g. 35, 42, 43, 44]. Schools and financial institutions closed in some towns and government services were terminated or reduced. Several towns were no longer able to provide business services such as the provision of farm equipment. The improvement of Karoo roads increased competition between towns, often to the detriment of smaller centres and caused leakages of purchasing power [42: 121]. The extent of such leakages for towns of the central Karoo was investigated [35: 116, 44: 73] and leakage rates in excess of 50 percent were recorded for some types of services.

Why did the larger Eastern Cape Karoo towns (Cradock, Graaff Reinet, Middelburg and Somerset East, see [38]) grow faster than smaller towns in the region? Le Clus [44: 8] and Vrey [35: 98] suggest the out-migration of whites and a reduction in the white birth rate played an important role in the regression of Karoo towns. Toerien [45], however, points out that in the period after 1951 when larger towns grew rapidly there was no obvious difference in the ratio of white people to non-white people in the larger towns versus the smaller towns of the Eastern Cape Karoo. He argues that it is unlikely that white out-migration is the single most important reason for the differential growth rates between the larger and smaller towns. The larger Karoo towns have diversified their economies and become less reliant on agriculture and more reliant on their own populations. The success of the larger centres is contrary to earlier predictions [38], however, the main reasons for their success still need to be determined [45].

The poverty and underdevelopment of black and coloured Karoo communities meant that traders in Karoo towns became ever more dependent on their growing black and coloured clientele [35: 98], a finding supported by Toerien [45]. Dependence on poorer clients was associated with a decline in variety, quantity and quality of retail facilities and services [35: 105].

The Eastern Cape Karoo towns are generally small and low in functional status, lack economic diversification and functional specialization, and have high levels of unemployment and high dependency ratios of economically inactive to economically active populations [42: 80]. Low levels of economic activity result in structural deficiencies, a lack of economic diversification, a general lack of employment opportunities and limited economic welfare that results in a lack of market potential, a lack of local sources of development capital and a loss of human skills and resources. By the 1970s the Cape midlands and Karoo area presented a picture of a static, and in certain instances retrogressive, economy when compared to the rest of the country [43: 1].

By 2008 urbanization dynamics in the Karoo mirrored the national trend towards enhanced urbanization since 1960 [38]. Nel and Hill [38] explain that rural depopulation seemed to match urban population growth. The degree to which activities clustered in larger, higher-order centres was distinctive in the Eastern Cape Karoo and paralleled business closure in smaller centres. The improved mobility of farmers favoured more distant, diversified and larger economic centres. The closure of schools, financial institutions, and government services in smaller centres contributed to the decline. The larger centres also diversified their economies, making them less reliant on the vulnerable rural economy and more reliant on their own larger populations. Services and tourism became more important. However, the decline of all Karoo towns anticipated in the 1970s did not materialize and the economic expansion of larger centres ran contrary to these predictions. The economic and population growth of the larger towns and the persistence of a strong rural economy, negated the perception that the Karoo is either in decline or is marginal [38]. They also suggest that the smaller service centres have survived economically through the emergence of a state welfare system. The valuable research of Nel and Hill [38] has, however, a significant weakness. It does not explain why some towns in the Eastern Cape Karoo grew economically and became large, while others remained stagnant or declined. For instance, one would expect the towns of Middelburg and Jansenville, founded respectively in 1852 and 1854 as so-called 'church towns' and both serving the needs of rural communities for religious services [39: 362], to be fairly similar in size. Yet by 2006 the population and number of enterprises in Middelburg were more than three times higher than those of Jansenville. More information is therefore needed to understand and predict the economic impacts of the new developments in the Karoo.

PART 2

Important Proportionalities in the Karoo

The presence of proportionalities (regularities) in any data set offers an ability to make predictions. Kahneman [25: 217-225] discusses the different outcomes between the predictions of experts and of simple algorithms. Surprisingly predictions based on simple statistics often outperform those of world-renowned experts. This suggests that if proportionalities are observed for Karoo towns, they could provide a sound basis to predict future economic, demographic and employment impacts. This ability is sorely needed in seeking win-win solutions for the utilization/conservation dilemmas of the Karoo.

Toerien and Seaman [27] investigated proportionalities in the enterprise structures of 125 South African towns. The magnitude of enterprise development in a town is: (i) a function of the population size of the town, and (ii) there are statistically significant relationships, and hence proportionalities, between the total number of enterprises in towns and some, if not all, of the enterprise numbers of different business sectors in towns [27]. The authors suggest that the implications of these proportionalities have far-reaching implications for rural development and job creation.

Toerien [46] and Toerien and Seaman [26, 28, 47] demonstrate proportionalities in the enterprise structures of some Karoo towns. Based on these Toerien and Seaman [29] propose a causal loop diagram (systems model) of the linkages between characteristics such as gross value added (GVA), population size, personal income, employment, entrepreneurship, enterprise numbers and their distribution. They argue that the model suggests that the money in a town, which is at least in part derived from the products produced or services rendered to local or external markets, serves to attract people to the town. As a result a dynamic balance develops between personal income and population numbers.

The population has specific needs and wants, which determine the total number of enterprises and the relative proportions of different enterprise types. In addition, the number of enterprises is in balance with the economic value added and the personal income of the town. The portion of the added economic value spent in a town together with other sources of money (e.g. government welfare payments, returns on personal savings and investments of residents, pensions paid to elderly residents, new investments by investors or entrepreneurs, etc.) determine the amount of money in a town.

Toerien and Seaman [29] also argue that in its vicious (negative) mode the model suggests that if the flow of money to a town diminishes, people will be less attracted to the town and current residents may even become dissatisfied and decide to move elsewhere. This reduces the total needs and wants that determine the total number of enterprises as well as the relative distribution of enterprise types. The reduction in money and enterprises will negatively influence the reputation of a town, and perhaps reduce the flow of investment and pension money to the town.

This systems model suggests that the regeneration of towns is primarily about generating new sources of money for towns [29]. This would attract more people and entrepreneurs, generate more needs and wants, and stimulate the development of new enterprises that would create more jobs and return more money by way of wages and salaries that could be spent in the towns. One can accept that the new developments in the Karoo (the SKA, solar power production and possibly shale gas production) will increase the inflow of money into local communities thereby generating the positive cycle described above, but to what extent and for how long? In this chapter I delve deeper into the interrelationships of the economic, demographic, personal income and entrepreneurial characteristics of eleven Karoo towns that occur in the area likely to be affected by shale gas exploitation (Figure 1).

A data set obtained from IHS Global Insight of the economic, demographic and personal income data for 2004 of eleven Karoo towns enabled the analyses. Enterprise numbers and structures for the selected towns were obtained from Toerien and Seaman [48]. Correlation coefficient and regression analyses were used to examine the proportionalities and Microsoft Excel™ software was employed for the calculations.

Relationships between Various Characteristics of Karoo Towns

The GVA, personal income, population numbers and enterprise numbers of the eleven towns, which ranged widely, were all highly significantly ($P < 0.01$) correlated with one another (Table 1). To quote Kahneman [25: 328]: "(T)his is a surprising finding that is trying to tell us something. What story is it trying to tell us?"

In unravelling the story one must keep in mind that correlations do not necessarily reflect causality. It would, however, be difficult to accept that these regularities do not reflect an underlying system in which causality plays a role. Occam's Razor [49] can be employed to guide us in the process. It states that among competing hypotheses, the one with the fewest assumptions should be selected, or in other words, to accept the simplest possible theoretical explanation for existing data.

Table 1. Correlations between and regression equations of a number of characteristics of the eleven Karoo towns

Relationship	Correlation*	Regression coefficient	Intercept
GVA-personal income	0.98	0.82	13.4
GVA-population	0.98	69.49	1630.0
GVA-enterprises	0.99	0.62	-11.3
Personal income-population	0.99	84.21	639.7
Personal income-enterprises	0.99	0.74	-18.6
Population-enterprises	0.98	0.0086	-21.1

*All correlations statistically significant ($P < 0.01$). The first mentioned characteristic was used as the independent variable in each comparison. All characteristics were for 2004 except for enterprises where the dataset of Toerien and Seaman [48] was used. GVA and personal income are expressed in constant (year 2000) rand million. Population and enterprises are expressed as total numbers per town.

Since the Industrial Revolution divergences in income have developed. For instance, Kennedy [50: 11] remarks: “The awful result, according to one calculation, was that whereas the British and Indian people had roughly similar per capita levels of industrialization at the onset of the Industrial Revolution (1750), India’s level was only one-hundredth of the United Kingdoms’ by 1900.” The divergence of national incomes and living standards that began with the Industrial Revolution continues to widen to the present [51: 319]. How do people react to such divergences?

There is historical evidence that people, usually poor, are attracted to places where they can make a better living. The attraction of the rich countries for the poor people of South America, Africa and South Asia has resulted in a series of fortifications ringing the modern West to protect it from invasion by poor societies [51: 373]. Even in the Karoo people are proportionally responding to where wealth is created (and not vice versa). It follows logically that GVA must be a significant driver of personal income.

The correlation between the total personal incomes and population numbers of the selected Karoo towns (Table 1) supports the notion that knowledge of high (relative to expectations) personal incomes in a town serves to attract people to it, or alternatively low personal incomes serve to repel people. The number of people in a town and their personal incomes help to determine the number of enterprises that can be 'carried' in the town.

Table 1 provides a direct as well as an indirect way of estimating the population impact of increases or decreases of GVA in Karoo towns: (i) the GVA-population relationship indicates that an increase/decrease of R1 million (in year 2000 rand) in GVA will add/shed just below 69.5 people, (ii) the GVA-personal income-population relationships suggest that an increase/ decrease of R1million (in year 2000 rand) will result in R820 thousand more/ less in total personal income, which will add/shed 69 persons. The two estimates are virtually identical and a value of 70 persons per GVA of R1 million (in year 2000 rand) can be used for predictions about Karoo towns.

The impact of changes on the number of enterprises can be estimated in three ways: (i) the GVA-enterprises relationship (Table 1) suggests that an increase/decrease of one enterprise will be associated with a positive/negative change of R1.61 million (year 2000 rand) in GVA, (ii) the GVA-personal income-enterprises relationships suggest that an increase/decrease of one enterprise will be associated with a positive/negative change of R1.35 million (year 2000 rand) in personal income, which equates to R1.65 million (year 2000 rand) of GVA, and (iii) the GVA-population-enterprises relationships suggest that an increase/decrease of one enterprise will be associated with a positive/negative population change of 116 persons, which equates to an increase/decrease of R1.67 million of GVA (year 2000 rand). The estimates are very similar and the average value of R1.64 million (year 2000 rand) per enterprise can be used to asses impacts on Karoo towns.

Delving Deeper into Gross Value Added Relationships

The fact that GVA appears to be a significant driver of events in Karoo towns raises the question as to whether the economic sectors contributing to GVA are also involved in regularities. I therefore examined their relationships with town populations (Table 2).

The GVAs contributed by three sectors, i.e. agriculture, mining and general services (partly) were not significantly ($P = 0.05$) correlated with the population sizes of the selected towns. All of the rest of the sectors were (Table 2). Larger towns have more economic value added in these sectors than smaller towns, and in proportion to their population sizes.

Table 2. The correlations and regression equations between the economic value added (GVA) (independent variables) by different business sectors in the selected Karoo towns and their population numbers (dependent variable)

Economic sector	Correlation	Regression coefficient	Intercept
Agriculture	0.55	378.3	-7635.1
Processing	0.82*	1801.1	6424.5
Factories	0.81*	1053.3	10147.1
Construction	0.89*	1637.7	5980.7
Mining	0.17	1520.8	14849.3
Tourism (partly)	0.84*	5955.9	2833.3
Financial services	0.92*	489.4	7836.8
Trade	0.91*	537.8	3700.9
Vehicle services	0.88*	2068.0	8877.1
General services (partly)	0.56	708.0	7901.1
Health services	0.87*	326.0	9128.3
Transport services	0.79*	298.6	10466.0
Real estate services	0.93*	627.7	4893.8
Public administration	0.96	416.8	5553.7
Education	0.95*	472.1	1746.8
Other business activities	0.70**	1212.6	10347.3
Other services activities	0.68**	333.2	8372.1

*Statistically significant at $P<0.01$, **Statistically significant at $P<0.05$.

To unravel these relationships further, I used principal component and cluster analyses, both multivariate statistical techniques [52]. The inputs for both analyses were normalized (as percentages) sector GVA contributions. Primer 6™ software was used for the calculations.

Principal Component Analysis (PCA)

A PCA is [53]:

> [A] statistical procedure that uses an orthogonal transformation to convert a set of observations of possibly correlated variables into a set of values of linearly uncorrelated variables called principal components. The number of principal components is less than or equal to the number of original variables. This transformation is defined in such a way that the

first principal component has the largest possible variance (that is, accounts for as much of the variability in the data as possible), and each succeeding component in turn has the highest variance possible under the constraint that it is orthogonal to (i.e., uncorrelated with) the preceding components. The principal components are orthogonal because they are the eigenvectors of the covariance matrix, which is symmetric.

In my analysis the agricultural sector contributes most (but negatively) to principal component 1 (Table 3). The public administration, transport services, financial services, other service activities and health services sectors contribute less and positively to this principal component (Table 3). Given the fact that this principal component explains 64.7 percent of the variance, these sectors are all important drivers of GVA differences in Karoo towns.

Table 3. The eigenvectors (coefficients in the linear combinations of variables making up the principal components) of the principal component analysis of the sector contributions to the gross domestic product (GDP)

Sector	Principal component 1	Principal component 2	Principal component 3
Agriculture	-0.906	0.024	-0.059
Processing	0.041	-0.033	-0.043
Factories	0.061	-0.030	0.026
Construction	0.050	-0.055	0.165
Mining	0.006	-0.016	-0.012
Tourism (partly)	0.015	-0.010	0.007
Financial services	0.170	0.347	-0.017
Trade	-0.090	0.155	0.034
Vehicle services	0.033	0.051	0.049
General services (partly)	-0.075	0.412	-0.080
Health services	0.145	-0.085	0.255
Transport services	0.175	0.134	0.039
Real estate services	-0.003	-0.002	0.358
Public administration	0.227	-0.093	0.059
Education	-0.026	-0.802	-0.094
Other business activities	0.029	-0.012	0.160
Other services activities	0.149	0.012	-0.851

The education sector contributes most (but negatively) to principal component 2 while the general services (partly) and financial services sectors are important positive contributors (Table 3). Given the fact that this principal component explains eleven percent of the variance, these sectors do play a part in the GVA differences of Karoo towns. In a similar fashion, the other service activities sector is the dominant (negative) contributor to principal component 3 and the real estate services, health services, other business activities and construction sectors contribute positively. This principal component explains 10.1 percent of the variance of GVA. The first three principal components explain a very large part (85.9 percent) of the variation of the GDP sectors, which reflects regularities in the underlying structure of how value is presently added in Karoo towns. To analyze this issue further I used cluster analysis.

Cluster Analysis

Cluster analysis is a method used to study the similarities within assemblages of entities. It is characterized by the fact that the comparison of two or more entities is based on the extent to which the entities share particular characteristics at comparable levels of abundance [52].

Similarity coefficients, in this case Euclidean distance, calculated between every pairing of towns, enable clustering of the towns into 'natural' groups that are generally more similar to each other than to towns in different groups. Seventeen normalized economic sectors were used to characterize the GVA patterns of the selected towns. Therefore, the GVA of each town can be thought of as being a point in a 17-dimensional Euclidean hyperspace where each dimension represents one of the GVA sectors. The distance between towns in this space reflects their similarities/dissimilarities. Towns with high similarities occupy points closer together in the hyperspace, and towns that are dissimilar are further away from each other. Three clusters of towns were identified at a Euclidean distance of 32 (Figure 2). The first cluster included Prince Albert, Richmond, Fraserburg and Carnarvon, the second cluster included Willowmore, Victoria West and Sutherland, and the third Graaff Reinet, Laingsburg, Middelburg and Beaufort West.

Clusters 1 and 2 are typified by a large contribution of agriculture to their total GVAs (Table 4). Donaldson and Marais [54: xi] remark that contributors to a recent book on small town geographies in South Africa hold different views about the relationships between small towns and the economic activities of their rural hinterlands. Some argue for increased linkages, others show that the historic linkages have deteriorated due to increasing global economic activity, changing transport modes and a decline in the agricultural sector. Agriculture is undoubtedly economically very important to the Karoo towns in clusters 1 and 2, but no longer so much to the Karoo towns of cluster 3.

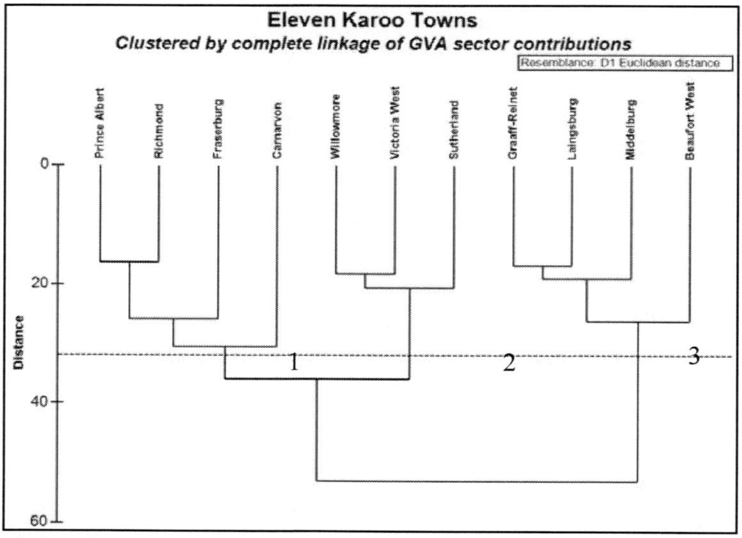

Figure 2. Dendrogram of the interrelationships of eleven Karoo towns based on their gross value added patterns. The broken line defines clusters 1, 2 and 3.

These towns have diversified their economies and the financial services, public administration, other services and factory sectors have become important GVA drivers (Table 4). These towns serve as examples of a post-productive realignment in South African towns [55: 28-30].

The SKA project and solar energy installations should enhance the trend to post-productive development in towns close to them. Mining, however, is presently not a big factor in the economic dynamics of the selected Karoo towns (Table 4). Should shale gas extraction proceed in the Karoo, this situation will change dramatically.

Table 4. Relative contributions of different economic sectors to gross value addition (GVA) of the three clusters of and eleven individual Karoo towns

	Cluster 1				Cluster 2			Cluster 3			
Sectors	Prince Albert	Richmond	Fraserburg	Carnarvon	Willowmore	Victoria West	Sutherland	Graaff Reinet	Laingsburg	Middelburg	Beaufort West
Agriculture	32.0	59.1	56.0	37.5	41.2	33.3	54.5	10.6	23.3	14.2	12.5
Processing	0.0	0.2	1.3	0.8	1.6	3.5	0.0	1.7	2.3	4.5	1.5
Factories	0.2	0.3	0.0	0.3	0.1	0.1	0.0	4.2	0.3	4.9	1.0
Construction	8.2	0.3	1.6	0.4	0.4	0.6	0.1	2.3	1.9	4.1	2.3
Mining	0.0	0.2	0.0	0.0	0.0	0.9	0.0	0.0	0.0	1.3	0.0
Tourism [partly]	1.5	0.8	0.0	0.2	1.4	1.3	0.0	0.8	2.9	0.8	0.5
Financial services	0.8	1.3	0.0	4.5	0.0	1.6	1.6	9.8	16.7	3.8	7.6
Trade	4.6	5.5	13.7	9.3	6.0	7.7	15.7	8.4	0.9	6.4	9.2
Vehicle services	1.0	0.4	0.4	0.0	0.0	0.8	0.0	2.3	3.3	0.2	1.6
General services [partly]	0.1	5.7	1.0	3.4	2.4	7.0	16.9	2.8	11.2	0.6	4.3
Health services	8.4	3.0	2.5	4.1	0.6	3.1	1.8	16.3	1.1	8.3	5.3
Transport services	2.7	0.0	1.0	1.2	1.7	1.7	0.0	5.1	7.1	5.0	15.9
Real estate services	11.1	4.1	15.1	2.8	4.2	2.7	2.6	7.4	5.3	5.2	7.6
Public administration	9.7	1.7	0.8	4.7	4.0	9.8	0.6	12.4	10.2	14.8	8.1
Education	12.3	15.6	5.7	6.5	27.8	16.7	4.5	11.7	4.1	11.5	8.4
Other business activities	7.5	0.0	0.8	0.0	0.0	1.7	0.6	1.1	1.7	0.0	3.4
Other services activities	0.0	1.8	0.0	24.4	8.5	7.6	0.9	2.9	7.7	14.6	11.0
Total GVA	100.0	100.0	100.0	100.0	100.0	100.0	100.0	100.0	100.0	100.0	100.0

Enterprise Proportionalities

It was shown earlier (Table 1) that the number of enterprises in Karoo towns are correlated with: (i) GVAs, (ii) population numbers, and (iii) total personal income. It was necessary to determine if proportionalities of the kind reported for a number of South African towns (e.g. [27]), are also present in Karoo towns. The reported proportionalities have two main aspects, namely (i) those based on enterprise numbers, and (ii) those based on enterprise structures.

Proportionalities based on enterprise numbers involve: (i) proportionalities of enterprise numbers in specific business sectors with town population sizes [27, 47], (ii) proportionalities of enterprise numbers in specific business sectors with the total enterprise numbers of towns [27, 47], (iii) functional business diversity relationships [48], and (iv) enterprise richness as a function of total enterprise numbers [48].

Enterprise structural proportionalities are based on normalized (as percentages) contributions of the enterprises in 19 different business sectors to the total number of enterprises in towns [26]. Multivariate statistical techniques such as cluster analysis and PCA have been used to elucidate and compare enterprise structures [26, 27, 28, 29, 48, 56] and have also been employed in this analysis.

For the analysis of enterprise structures the investigation was not limited by the non-availability of economic data at the level of individual towns. Therefore a much larger group (42 towns) of Karoo towns (Table 5), that includes the eleven towns used earlier, could be analyzed. Enterprise numbers and structures for the selected towns were obtained according to Toerien and Seaman [26].

Enterprise Numbers versus Population Numbers of Towns

Correlation and regression analyses between town population numbers (independent variable) and the number of enterprises in 19 different business sectors as well as the the total number of enterprises (independent variables) are presented in Table 6.

There is an extraordinarily strong regularity between the population numbers of and the total number of enterprises in Karoo towns and the regression equation explains more than 95 percent of the variance (Table 6). Even when the population relationships with different business sectors are considered, the regression equations for six business sectors (trade, vehicle, personal, general, financial and construction services) explain more than 80 percent of their variances. For another ten business sectors (health, telecommunication, tourism and hospitality, professional, agricultural products and services, real estate, transport and earthworks) between 65 and 79 percent of the variances are explained. For only four sectors (news and advertising, factories, processors and mining) are less than 50 percent of their variances explained. These strong regularities probably indicate a very dynamic relationship between populations and numbers of enterprises, suggesting the absence of time lags in the responses of enterprises enterprises (and entrepreneurs) to changes in population.

Table 5. The Karoo towns, their populations and enterprises used to examine enterprise proportionalities

Town	Population (2011)	Enterprises	Year	Town	Population (2011)	Enterprises	Year
Aberdeen	7162	46	2010/11	Murraysburg	5069	21	2006/07
Beaufort West	34085	353	2008/09	Nieu-Bethesda	1540	21	2006/07
Brandvlei	2859	22	2009/10	Nieuwoudtville	2093	30	2009/10
Britstown	5145	27	2008/09	Noupoort	7848	30	2010/11
Calvinia	9680	110	2008/09	Pearston	4516	20	2010/11
Carnarvon	6612	58	2008/09	Petrusville	5211	17	2007/08
Colesberg	16869	144	2006/07	Philipstown	3365	15	2007/08
Cradock	36671	294	2010/11	Prieska	14246	108	2008/09
De Aar	29990	223	2008/09	Prince Albert	7054	83	2009/10
Fraserburg	3029	33	2008/09	Richmond	5122	30	2002/03
Graaff Reinet	35672	347	2010/11	Somerset East	18825	194	2010/11
Hanover	4594	22	2008/09	Steynsburg	7212	38	2010/11
Hofmeyr	3680	17	2010/11	Steytlerville	4017	34	2010/11
Hopetown	10259	70	2008/09	Strydenburg	2987	17	2009/10
Jansenville	5612	49	2010/11	Sutherland	2836	35	2007/08
Kenhardt	4843	29	2009/10	Vanwyksvlei	1721	8	2009/10
Klipplaat	2967	13	2008/09	Venterstad	4989	18	2010/11
Laingsburg	5667	56	2008/09	Victoria West	8254	74	2000/01
Loeriesfontein	2744	29	2009/10	Vosburg	1259	9	2013/14
Loxton	1053	7	2009/10	Williston	3368	26	2000/01
Middelburg	18681	161	2010/11	Willowmore	7678	53	2010/11

Year refers to the year of the telephone directory used in the analysis of individual towns.

The main implication of this statement is startling: there appears to be no lack of entrepreneurs in Karoo towns, at least in those business sectors for which a very large part of their variances are explained (see Table 6).

This finding flies directly in the face of planning documents such as the National Development Plan [57: 85-86] where general entrepreneurship training is proposed for youth development. If such trained persons were to enter the above mentioned sectors in Karoo towns they would likely encounter very strong competition and have a high likelihood of failure.

Table 6. The correlations and regression equations of town population numbers (independent variable) and the total as well as business sector enterprise numbers (dependent variables) of 42 Karoo towns

Business Sector	Correlation	Variance explained (%)	Regression coefficient	Intercept
All enterprises	0.98	95.5	0.009219	-9.4
Trade Services	0.97	93.3	0.002148	-2.2
Vehicle Services	0.96	92.6	0.000976	-2.2
Personal Services	0.94	88.0	0.000714	-0.8
General Services	0.94	87.8	0.000472	-1.0
Financial Services	0.92	84.0	0.000609	-0.1
Construction Services	0.91	83.4	0.000602	-2.5
Health Services	0.89	78.9	0.000593	0.2
Telecommunication Services	0.88	77.8	0.000104	-0.4
Tourism and Hospitality	0.85	72.8	0.001223	0.1
Professional Services	0.85	72.5	0.000321	-0.7
Agricultural Products and Services	0.83	69.7	0.000503	1.1
Real Estate Services	0.83	69.1	0.000196	-0.5
Engineering and Technical Services	0.82	67.1	0.000211	-0.2
Transport and Earthworks Services	0.81	65.6	0.000226	-0.3
Legal Services	0.81	65.3	0.000147	0.0
News and Advertising Services	0.68	46.4	0.000030	-0.1
Factories	0.68	45.9	0.000042	-0.1
Processors	0.61	37.1	0.000106	0.1
Mining Services	-0.05	0.3	-0.000002	0.2
		Total:	0.009219	

The unit of the regression coefficient is enterprises per person. All correlations are statistically significant ($P < 0.01$) except that of the mining sector.

A second implication of these regularities is that the responses of Karoo towns to external influences such as increases/decreases in population numbers as a result of money flows to local economies are quite predictable, not only for total enterprise numbers but also for how these would be distributed among different business sectors.

The inverse of the regression coefficient for the regression equation (see Table 6) between town populations and total number of enterprises for the 42 Karoo towns provides an estimate of how many persons are generally needed to support a single enterprise in Karoo towns. This number is 108 persons. It compares very well with the inverse of the population-enterprises relationship presented in Table 1, which translates to 116 persons per enterprise. A number of 112 persons per enterprise can, therefore, be used for predictions about enterprise development in Karoo towns. In addition the inverse of the regression coefficients for the top 15 business sectors (Table 6) can be used in a similar fashion to make predictions of how these business sectors of Karoo towns would respond to population changes as a result of new developments in the Karoo.

A third implication of these findings is that because of the strong regularities between populations and total enterprises and between populations and sector enterprises, there must be regularities between sector enterprises and the total enterprises of Karoo towns. An analysis of this issue follows.

The Relationships between Total Enterprises and Sector Enterprises

These relationships were examined by normalizing the number of enterprises of each of the 19 different business sectors as a percentage of the total enterprises of each Karoo town. The normalized data provided assessments of the enterprise structures of towns and allowed comparisons between them irrespective of their sizes [26]. To assess similarities/ dissimilarities in the enterprise structures of the 42 towns, a cluster analysis was done using the normalized structures as input and once again Euclidean distance as the similarity index. Primer 6™ software was used for the analysis.

Six clusters of towns were identified at an Euclidean distance of 34: two larger ones, four small ones and an outlier (Nieu-Bethesda) (Figure 3). Cluster 1 contains nine towns, all with strong connections to tourism. Cluster 2 with 18 towns contains a mixture of large and small towns while clusters 3, 4, 5, and 6 respectively contain only 4, 3, 4 and 3 towns (Figure 3). To understand the differences between the clusters the total number of enterprises in each business sector for all of the towns in a cluster were expressed as a percentage of the total number of enterprises in each cluster (Table 7). Cluster 2 has some enterprises in all business sectors and the enterprise structure is fairly balanced. The towns of this cluster are possibly not exposed to undue risk of changes in a single or a few business sectors, e.g. when compared to the outlier (Nieu-Bethesda) with its absolute dependence on a single sector, i.e. the tourism and hospitality sector (see Table 7).

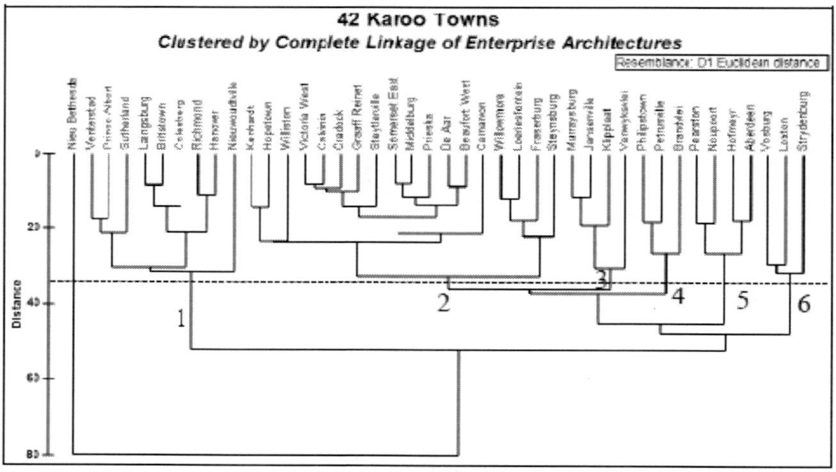

Figure 3. Dendrogram of the interrelationships of 42 Karoo towns based on their enterprise structures. The broken line defines clusters 1, 2, 3, 4, 5 and 6 as well as an outlier (Nieu-Bethesda).

Clusters 3 and 6 are strong in the agricultural products and services sector but cluster 3 is also strong in the trade and real estate sectors while cluster 6 is weak in these sectors. However, cluster 6 is the strongest cluster in the processor sector (adding value to primary products) and the engineering and technical services sectors.

Cluster 4 is the strongest in vehicle, financial and general services and its trade sector is reasonably strong.

Cluster 5 is strong in the trade services, personal services and health services sectors but is, given the rural nature of the towns, surprisingly weak in the agricultural products and services sector.

The use of cluster analysis enables the understanding that, despite the extent of proportionalities in enterprise numbers considered earlier, there are also differences in the enterprise structures of Karoo towns. These differences can be interpreted on the basis of a few clusters of towns as shown in Table 7. A deeper understanding of the reasons why there are clusters of Karoo towns based on their enterprise structures still needs to be developed.

Table 7. The average sector enterprise composition (%) of the clusters of Karoo towns

Business sector	Cluster 1	Cluster 2	Cluster 3	Cluster 4	Cluster 5	Cluster 6	Outlier
Agricultural Products and Services	7.9	7.0	23.1	7.4	5.3	21.2	9.5
Processors	2.0	1.3	1.1	5.6	0.9	9.1	0.0
Factories	0.2	0.4	0.0	0.0	0.9	0.0	0.0
Construction Services	1.1	4.9	0.0	1.9	1.8	0.0	0.0
Mining	0.0	0.2	1.1	1.9	0.0	0.0	0.0
Tourism and Hospitality Services	27.6	12.6	11.0	11.1	12.4	12.1	71.4
Engineering and Technical Services	2.0	2.6	0.0	0.0	0.9	9.1	0.0
Financial Services	5.6	7.9	4.4	14.8	4.4	3.0	0.0
Legal Services	2.2	1.9	1.1	0.0	0.9	0.0	0.0
Telecommunication Services	0.4	0.9	0.0	0.0	0.0	0.0	0.0
News and Advertising Services	0.0	0.3	0.0	0.0	0.0	0.0	0.0
Trade Services	19.8	24.0	27.5	22.2	26.5	9.1	9.5
Vehicle Services	8.5	9.2	3.3	16.7	4.4	12.1	0.0
General Services	3.6	4.7	4.4	9.3	2.7	0.0	0.0
Professional Services	3.4	2.9	4.4	0.0	2.7	0.0	0.0
Personal Services	6.7	7.7	4.4	0.0	15.9	9.1	4.8
Health Services	5.6	7.3	9.9	5.6	18.6	12.1	4.8
Transport and Earthworks Services	0.9	2.7	2.2	3.7	1.8	0.0	0.0
Real Estate Services	2.2	1.6	2.2	0.0	0.0	3.0	0.0
Total	100.0	100.0	100.0	100.0	100.0	100.0	100.0

In other studies, clustering has also enabled the detection of a phenomenon called proportionality-in-proportionality [46], i.e. a situation where clusters of towns had stronger proportionalities between the number of tourism and hospitality enterprises and the total number of enterprises than for the whole group of towns that were studied. Proportionality-in-proportionality in the tourism and hospitality sector was also observed in this study (Table 8). The variances explained by the regression equations for clusters 1 and 2 were quite a bit higher than that explained by the regression equation for all 42 towns.

There were also distinct differences in the regression coefficients for clusters 1 and 2 and the high regression coefficient for cluster 1 was effectively masked when all towns were analysed together. Larger Karoo towns in general have more tourism and hospitality enterprises but the towns in cluster 1 are especially strong in this sector.

Functional Business Diversity in Karoo Towns

Biological diversity is an important attribute of natural ecosystems and higher diversity is often, but not always, thought to be associated with ecosystem stability [58: 1457-1458].

Does enterprise diversity play a role in the business stability of towns? Toerien and Seaman [56] used an enterprise diversity index, a functional index based on the relative presence of 19 different business sectors in South African towns, to examine functional differences between towns in water abundant and water poor areas of South Africa. Their study suggests that functional business diversity is important in South African towns and it was, therefore, also examined for the 42 Karoo towns. It is assumed that each of the 19 business sectors represents an important function delivered in Karoo towns. The functional diversity index (FDI) of each town was estimated as follows:

$$FDI_{town} = (\text{No. of business sectors with at least one enterprise x } 100)/19$$

An important feature was revealed when the FDIs were plotted against the total number of enterprises per town (Figure 4): at about 100 enterprises there is a tipping point (see arrow in Figure 4), below this number there was a sharp and consistent drop in the FDI.

Above this number most of the business sectors (with the exception of the mining sector) were present in Karoo towns. Karoo towns with fewer than 100 enterprises might thus be at risk of business instability. The average number of enterprises in the 42 Karoo towns is 71.2 and the median number is 31.5. This indicates a very skewed size distribution for Karoo towns, with a few large towns and many small ones which might all be at risk of losing business sectors if their GVAs decline.

Table 8. Regression equations of the total number of enterprises (independent variable) and of the enterprise numbers of the tourism and hospitality sectors (dependent variable) of all 42 Karoo towns as well as the towns of clusters 1 and 2

Grouping	Correlation	Variance explained (%)	Regression coefficient	Intercept
All towns	0.91	82.8	0.138	1.0
Cluster 1	1.00	99.0	0.274	0.1
Cluster 2	0.95	91.1	0.144	-2.2

For instance Figure 4 indicates that at the median size (i.e. 31.5 enterprises per town), representing half of the Karoo towns, only about 50 percent of the business sectors in towns will have at least one enterprise.

Most Karoo towns will, therefore, benefit considerably if new developments bring more money into their economies and their enterprise structures and business services will expand. Conversely, if such developments terminate, the process will go the other way.

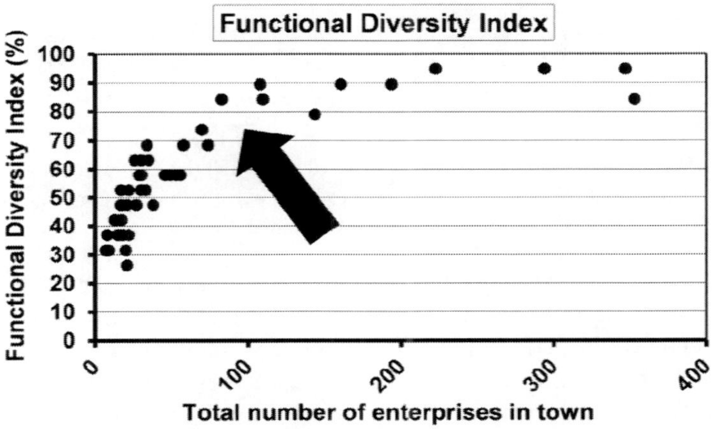

Figure 4. The functional diversity index (FDI) of 42 Karoo towns as a function of the size of the towns as measured by the total number of enterprises per town.

Enterprise Richness of Karoo Towns

The enterprise richness, i.e. the number of different business types, in a range of South African towns has been investigated [48]. A statistically significant power law (log-log) describes the relationship between enterprise richness and the size of towns (when expressed in terms of their total enterprise numbers). The enterprise richness of the 42 Karoo towns was examined in a similar fashion using the method of Toerien and Seaman [48].

The enterprise richness of the 42 Karoo towns is also highly significantly ($P< 0.01$) correlated in a power law with the size of the enterprise structures of the towns (Figure 5). The enterprise richness of Karoo towns, therefore, varies predictably over their entire size range (7 – 353) of enterprises.

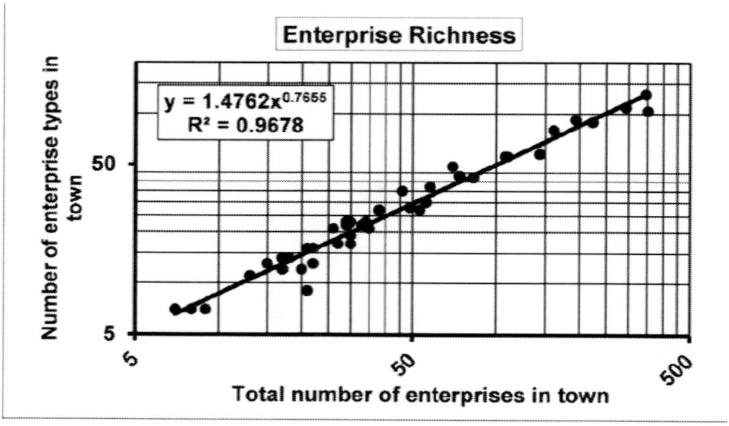

Figure 5. The log-log relationship of enterprise richness and the total number of enterprises in 42 Karoo towns. The relationship is statistically highly significant ($P < 0.01$) and almost 97 percent of the variance is explained.

Toerien and Seaman [48] interpreted the power law relationship in terms of how entrepreneurial space is created or destroyed as towns grow or regress. The entrepreneurial space in towns can be divided into two types: (i) business areas in which there are already existing entrepreneurs, and (ii) business areas in which there are no entrepreneurs and which require entrepreneurs to pursue business opportunities that have not yet been exploited (i.e. 'new entrepreneurs'). The power law enables the calculation of how new and existing entrepreneurial opportunities increase as functions of the magnitude of total enterprises (Figure 6).

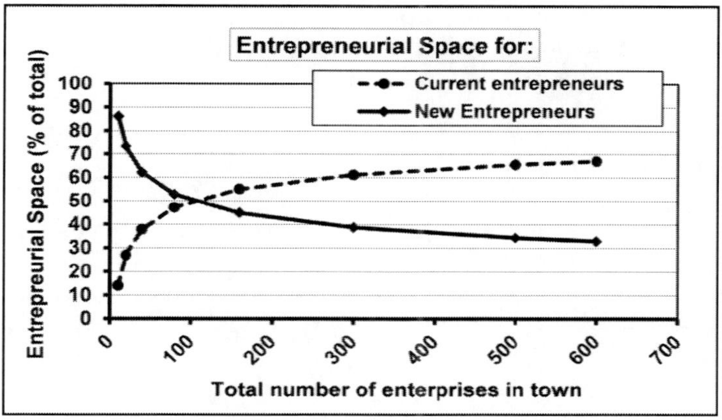

Figure 6. The entrepreneurial spaces for existing and new entrepreneurs as functions of the number of enterprises of Karoo towns. The entrepreneurial spaces were calculated from the differences between total enterprises and enterprise richness numbers through the use of the log-log relationship of Figure 5.

Below 100 enterprises in Karoo towns, enterprise growth is more associated with new entrepreneurs than existing entrepreneurs (Figure 6). At about 100 enterprises both types of entrepreneurs contribute equally to entrepreneurship but beyond 100 enterprises per town the opportunities for new entrepreneurship converge slowly on a contribution of just more than 30 percent of all business opportunities. Existing entrepreneurship converges on a value of just below 70 percent in large Karoo towns. In smaller Karoo towns new entrepreneurship is, therefore, more important than in larger Karoo towns; however, new and existing entrepreneurship are important in both small and large towns.

This can be illustrated by considering two scenarios involving an increase of 50 enterprises: (i) an increase from 20 to 70 enterprises in a small Karoo town, and (ii) an increase from 220 to 270 enterprises in a large Karoo town (Table 9).

In scenario 1 enterprise types increase by 23 and in scenario 2 by 15. New entrepreneurship is, thus, more crucial in small towns, but is also a factor in large towns. In scenario 1, duplication of existing business opportunities amounts to only 27 but in scenario 2 to 35. Duplication of existing business opportunities is, therefore, more pronounced in large towns than small ones (Table 9).

Table 9. Analysis of increases in enterprise types and duplication of existing business opportunities in small (scenario 1) and large (scenario 2) Karoo towns

Scenario	Total enterprises	Enterprise Types	Duplication of existing business opportunities
1	20	15	5
	70	38	32
2	220	92	128
	270	107	163

PART 3

Proportionalities in the Karoo and the Future

Quoting Kahneman [25: 328] again: what story about Karoo towns do the proportionalities that have been reported here, tell us? First, let us tally the proportionalities:

- GVA correlates positively with: town population size, total personal income and enterprise numbers in towns (Table 1).
- Total personal income also correlates positively with: town population size and enterprise numbers in towns (Table 1).
- Town population size also correlates positively with the number of enterprises in towns (Table 1, and also based on Table 5).
- GVAs of most business sectors correlate positively with the town population sizes (Table 2).
- Total enterprise numbers of Karoo towns correlate positively with the enterprise numbers of all business sectors except the mining sector (Table 6).
- Karoo towns with fewer than 100 enterprises (i.e. most of the Karoo towns) are in continuous danger of losing entire business sectors and their services (Figure 4).
- Enterprise richness has a statistically significant power law (log-log) relationship with the total enterprises of Karoo towns (Figure 5), which defines the expansion/contraction of entrepreneurial spaces for new and existing entrepreneurs (Figure 6).

Another of the consequences of order in Karoo towns is that there are discernible similarities in economic and enterprise structures:

- Three clusters based on GVA sector patterns among eleven Karoo towns (Figure 2).
- Five clusters and an outlier among 42 Karoo towns based on their enterprise structures (Figure 3 and differences summarized in Table 7).
- The detection of a proportionality-in-proportionality phenomenon in the tourism and hospitality sector of the 42 towns (Table 8).

There is no doubt that significant order, or in other words a strong underlying structure, is present in the business and enterprise dynamics of Karoo towns and the factors that drive these. Dawes [59] wrote and talked about the utility of using linear models in socially important decisions and debunked criticisms against their use. Kahneman [25: 222-223] describes how simple algorithms provided better predictions in many fields than experts do. For the purposes of this chapter, I worked with the proposition that the proportionalities described above can be used to make predictions about the potential economic, demographic and entrepreneurial impacts of new developments in the Karoo.

However, there is little quantitative information in South Africa about the impacts of the SKA and solar power generation projects. I have had informal discussions with business persons in the Karoo town of De Aar (solar power generation project) and the town of Carnarvon (SKA project) and was assured that these activities have benefited traders such as general dealers and butcheries as well as accommodation establishments and restaurants.

More is known about the impacts of the shale oil and gas extraction in the US. The shale revolution in the US is not just a temporary blip but a true game changer for the US economy and its impact is only in the early stages [60]. Nyquist and Lund [60] projected in 2013 that shale energy could by 2020 boost the US annual GDP by 2 to 4 percent ($380 billion to $690 billion). During 2008 the Marcellus gas industry in Pennsylvania generated $2.3 billion in total value, added more than 29,000 jobs, and $240 million in state and local taxes [16]. The authors [16] expect the economic output in 2009 to top $3.8 billion, state and local tax revenues to be more than $400 million and total job creation to exceed 48,000.

Kinnaman [18], however, points out that the former report has major shortcomings, in particular some questionable assumptions. The decline in the world oil price, however, is partly a result of shale oil extraction in the US [2]. States where shale gas and oil are being extracted benefit from the activity.

To form some ideas of the potential economic impacts that could be associated with the exploitation of shale gas in the Karoo it was necessary to use US information. Krauss [21] indicates that each drilling rig represents about 100 jobs, from roughneck field hands to maintenance workers. Robust statistical support was found for the hypothesis that changes in drilling for oil and gas as captured by rig counts, do have an economically meaningful and positive impact on employment [61]. It is estimated that an additional rig results in the creation of 37 jobs immediately and 224 jobs in the long run, but state that the multipliers could be bigger [61].

Estimates were provided in February 2015 of operating drill rig counts on the Eagle Ford Shale in Texas [62]. These varied from 31 rigs in Karnes County to some counties without any rigs. The median number of rigs in counties with rigs was eleven. The counties with rigs had a median area of 2,795 km^2, translating to 251.2 km^2/drilling rig.

Three of the South African local municipalities in whose domains shale gas exploitation might take place (if it is approved and economically viable) are Camdeboo, Beaufort West and Emthanjeni. Their municipal areas are respectively 12,422, 21,917 and 13,472 km^2. Using the median number of km^2/rig in counties on the Eagle Ford Shale (see above) this translates into 49, 87 and 54 rigs/municipality for the respective municipalities. This suggests that shale gas exploitation in the Karoo could mean that many rigs might operate simultaneously in a single local municipal area and employ a lot of people. Not all of the workers would be of local origin and nor would all of their salaries/wages be spent locally. They would nevertheless have to be housed, fed and entertained locally.

If we conservatively use a figure of 30 rigs/municipality, each with 50 workers, it means that at any point in time there could be 1,500 workers involved in shale gas extraction in a local municipality.

To develop a feel for the magnitude of the potential economic impact of such a limited scenario (just the economic impact of accommodation and meals) a 'back-of-an-envelope' calculation is necessary. Assuming that just accommodation and meals would cost R300 per person per day, an additional R13.5 million per month or R162 million per annum could be injected into the local economies of some Karoo towns for the duration of the shale gas exploitation, which might last some years. Expressed in year 2000 rand this equates to R62.1 million, which is equal to large parts of the GVAs (expressed in year 2000 rand) of the eleven Karoo towns analysed here, especially the smaller ones (Table 10).

The impacts of a monetary injection of this magnitude in terms of the demographic, personal income and entrepreneurial characteristics of a typical Karoo town can now be predicted from the proportionalities reported earlier. Based on the value of the GVA-population and the GVA-personal income-population relationships of 70 persons more/less per R1 million (in year 2000 rand) change, the populations of towns with such extraction activities in their vicinities could grow by about 4,300 persons. An influx of so many people would have a major impact on small Karoo towns, not only in terms of economic expansion but also in terms of infrastructure requirements and the requirement to provide adequate municipal services.

An increase of 4,300 persons in a typical Karoo town would result in between 37 (from the inverse of population-enterprises regression coefficient in Table 1) and 41 (from the inverse of regression coefficient of all enterprises in Table 6) additional enterprises, which would grow the enterprise structures of most of the Karoo towns significantly (compare with Table 5).

Table 10. A comparison of R62.1 million (year 2000 rand) to the GVAs (year 2000 rand) of eleven Karoo towns

Town	Percentage of GVA
Richmond	97.8
Prince Albert	82.6
Laingsburg	82.0
Fraserburg	75.1
Sutherland	62.4
Willowmore	59.7
Carnarvon	48.8
Victoria West	38.2
Middelburg	23.2
Graaff Reinet	12.4
Beaufort West	11.6

Such enterprise growth would move the enterprise count of the median town from 32 to between 68 and 72. This would also increase enterprise types from 21 to between 37 and 39 (calculated from the equations in Figure 5).

There would thus be significant new as well as existing entrepreneurial opportunities. Some of these would be associated with services in new business sectors.

Figure 4 suggests that an enterprise increase of that magnitude would move the FDI number of the median town from about 50 (activity in eight to nine business sectors) to 70 (13 business sectors present). This would be to the benefit of the population of the town because services that were not available before would now be present.

CONCLUSION

De Wit [8] recently commented on the debate about the possible use of fracking to extract gas from shales in South Africa. He points out that exploration is an inherent part of a process to learn about the benefits and detriments of local gas extraction, and that a collective learning process should be pursued. Disciplines/areas that should be included in the learning process are: economics, sociology, geology, hydrology, energy, engineering and environmental sciences. He suggested that the collective learning process should be transparent in order to overcome the negative spiral of a polarization of opinions about fracking in South Africa.

As pointed out earlier the debate in South Africa about the fracking of Karoo shales is really about a conservation/utilization dilemma. This dilemma has been faced in the Karoo in many different guises over its history. Utilization rather than conservation started in the Karoo when Khoikhoi herders moved in with their herds [31] and was expanded when trekboers settled in the Karoo to farm [32]. The introduction of merino sheep and wool production later became important economic drivers [33: 66], aided by wind pump technology to sink boreholes and extract groundwater for livestock use (ironically wind pumps are now considered to be Karoo icons). Utilization of the Karoo continued with the founding of the Karoo towns [39: 361-362]. At first rough tracks in the Karoo facilitated the transport of passengers and goods [32: 59], but in time roads and railways were developed [63].

Modern travellers during their travels through the Karoo often see lines of pylons forming part of the national electricity grid. The Karoo landscape has, therefore, over its history been subjected to many different uses. The question is now if the new uses of the Karoo (the SKA project, solar power generation projects and possible fracking of shales to extract gas) are acceptable uses of the Karoo landscape and if so, where.

Hampden-Turner [7: 19] suggests that both 'horns' of a dilemma (which he called unbalanced excesses) should be avoided. In the case of the debate on the fracking of Karoo shales the horns of the dilemma are: (i) total extraction of gas without concern about detrimental impacts, and (ii) a total ban on gas extraction without consideration of the benefits that might arise. Hampden-Turner [7: 21-23] advises that win-win solutions should be sought. Obviously such solutions would only arise if the advice of De Wit [8] is heeded, and this requires the availability of many kinds of information.

This study contributes information to the shale debate. It demonstrates that: (i) there is a strong underlying structure and order in the economic, demographic and entrepreneurial relationships of Karoo towns, (ii) the quantified relationships have predictive power, and (iii) cluster and other analyses have shown that there are both important similarities in and differences between the enterprise structures of clusters of Karoo towns.

My predictions are based on: (i) the relationships that have been presented here, (ii) use of the characteristics of the so-called median town, and (iii) assumptions developed from shale gas and oil extraction in the US about the number of borehole rigs employed and manpower used. As more information becomes available locally about the actual conditions under which gas extraction (should it be approved) will take place, these predictions can be improved and expanded by taking the characteristics of clusters of towns as well as individual towns into account. Given the unemployment and poverty situation in South Africa (also in the Karoo) the current and future predictions are important and a necessary part of the process to develop a balanced view of the benefits and detriments of shale gas extraction in the Karoo. They should be used.

The exploration phase for shale gas in the Karoo is expected to last some years and this time window should be utilized to gather objective facts. The advice of De Wit [8] should, therefore, be heeded, i.e. the mobilization of a wide range of disciplines to move South Africa towards a win-win situation as far as shale gas extraction in the Karoo is concerned. This task will be extremely difficult given the level of the current debate and evidence of some 'paradigm paralysis' as illustrated by demands for 'one horn' solutions, i.e. no shale gas extraction in the Karoo as opposed to total utilization of the Karoo for the extraction of shale gas.

The above means that academics and scientists should provide objective facts as demanded by their professional training. This contribution supplies information about a part of the puzzle that must be solved, i.e. the development of an ability to predict the demographic and entrepreneurial consequences of an increased inflow of money in the local economies of the Karoo towns.

Finally, the existence of a conservation/utilization dilemma in the potential extraction of shale gas in the Karoo should be accepted by persons and organizations involved. Only cooperation across many disciplines and areas [8] can help find a win-win solution. South Africa and the Karoo need that.

ACKNOWLEDGMENTS

Financial and other support by the Centre for Environmental Management, University of the Free State, Bloemfontein and technical assistance by Frank Sokolic, Linda Retief, Marie Toerien and Estelle Zeelie are gratefully acknowledged.

REFERENCES

[1] Atkinson, D. and Marais, L. (2007). *District socio-economic profile and development plans. Report.* Bloemfontein: Centre for Development Support, University of the Free State.

[2] Nocera, J. (2014). *Shale and the falling price of oil.* New York Times, December 22, 2014.

[3] US Energy Information Administration. (2013). *Technically recoverable shale oil and shale gas resources: An assessment of 137 shale formations in 41 countries outside the United States.* Washington, DC: US Department of Energy.

[4] De Wit, M. J. (2011). The great shale debate in the Karoo. *South African Journal of Science* 107(7/8), Art. #791, 9 pages. doi:10.4102/sajs.v107i7/8.791.

[5] *SouthAfrica.info*. (2014). SA to regulate, monitor shale gas exploration. Accessed on January 8, 2015 at: http://www.southafrica.info/about/sustainable/shale-gas-280814.htm#.VK4iuSuUdqU.

[6] De Jongh, M. (2002). No fixed abode: The poorest of the poor and elusive identities in rural South Africa. *Journal of Southern African Studies*, 28, 441-460.

[7] Hampden-Turner, C. (1990). *Creating corporate culture: From discord to harmony*. Reading, MA: Addison-Wesley Publishing Company.

[8] De Wit, M. (2015). Breking en die ekonomie: Gaan die land 'n wins kan maak? [In Afrikaans] *Litnet* 2015-02-13. Accessed at: http://www.litnet.co.za/Article/breking-en-die-ekonomie-gaan-die-land-n-wins-kan-maak.

[9] *Engineering News*. (2014). Karoo fracking could start by mid-2016. Accessed at: http://www.engineeringnews.co.za/article/karoo-fracking-could-start-by-mid-2016-2014-10-28.

[10] SKA Africa. (2015). *Meerkat*. Accessed at: http://www.ska.ac.za/meerkat/index.php.

[11] Brits, E. (2015). *Teleskoop-bou is nie huisbou nie. [In Afrikaans]*. Cape Town: Die Burger, (p. 15) Friday, 23 January 2015.

[12] *Wikipedia* [2015a] Solar power in South Africa. Accessed on January 22, 2015 at http://en.wikipedia.org/wiki/Solar_power_in_South_Africa.

[13] Deutch, J. (2011). The good news about gas: The natural gas revolution and its consequences. *Foreign Affairs*, 90, 82-93.

[14] Morse, E. L. *Energy* 2020: North America, the new Middle East? Accessed at http://csis.org/files/attachments/120411_gsf_MORSE_ENERGY_2020_North_America_the_New_Middle_East.pdf.

[15] IHS Global Insight. (2009). Measuring the Economic and Energy Impacts of Proposals to Regulate Hydraulic Fracturing. *Report for the American Petroleum Institute*, Lexington, MA: IHS Global Insight.

[16] Considine, T., Watson, R., Entler, R., and Sparks, J. (2009). *An emerging giant: Prospects and economic impacts of developing the Marcellus Shale Natural Gas Play*. State College, PA: College of Earth and Mineral Sciences, Pennsylvania State University.

[17] Considine, T., Watson, R. and Blumsack, S. (2010). *The economic impacts of the Pennsylvania Marcellus Shale Gas Play: An update*. Accessed at: http://marcelluscoalition.org/wpcontent/uploads/2010/05/PA-Marcellus-Updated-Economic-Impacts-5.24.10.3.pdf.

[18] Kinnaman, T. C. (2010). The economic impact of shale gas extraction: A review of existing studies. *Other Faculty Research and Publications, Paper 5*. Bucknell University. Accessed at: http://digitalcommons.bucknell.edu/fac_pubs/5.

[19] Hammond, G. W. (2010). *West Virginia economic outlook 2011*. Morgantown, WV: Bureau of Business and Economic Research, West Virginia University.

[20] Weber, J. G. (2012). The effects of a natural gas boom on employment and income in Colorado, Texas, and Wyoming, *Energy Economics*, 34, 1580-1588.
[21] Krauss, C. (2011). Shale boom in Texas could increase US oil output, *New York Times*, May 27, 2011.
[22] Weisman, J. (2012). In Western Pennsylvania, an energy boom not visibly stifled. *New York Times*, June 20, 2012.
[23] Healy, J. (2012). Western Colorado struggles as energy jobs fade. *New York Times*, June 28, 2012.
[24] Schwartz, N. D. (2014). Boom in energy spurs industry in the Rust Belt. *New York Times*, September 8, 2014.
[25] Kahneman, D. (2011). *Think, fast and slow*. Kindle Edition, Penquin.
[26] Toerien, D. F. and Seaman, M. T. (2010). The enterprise ecology of towns in the Karoo, South Africa. *South African Journal of Science* 106, 24-33.
[27] Toerien, D. F. and Seaman, M. T. (2012a). Proportionality in enterprise development of South African towns. *South African Journal of Science* 108, 38-47.
[28] Toerien, D. F. and Seaman, M. T. (2012b). Evidence of island effects in South African enterprise ecosystems. In: A. Mahamane (Ed.), *The functioning of ecosystems* (pp. 229-248). Rijeka: Intech.
[29] Toerien, D. F. and Seaman, M. T. (2012c). *Paradoxes, the tyranny of structures and enterprise development in South African towns*. Presented at Towards Carnegie III Conference, 3-7 September 2012, Cape Town.
[30] Cowling, R. M. (1986). A description of the Karoo Biome Project. *South African National Scientific Programmes Report no. 122*. Pretoria: Foundation for Research Development.
[31] Boonzaier, E., Malherbe, C., Smith, A., and Berens, P. (1996). *The Cape herders: A history of the Khoikhoi of Southern Africa*. Athens, OH: Ohio University Press.
[32] Elphick, R. (1979) The Khoisan to c. 1770. In: R. Elphick and H. Giliomee (Eds.). *The shaping of South African society*, 1652-1820. (pp. 3-40), Cape Town, Maskew Miller Longman.
[33] Wickins, P. L. (1983). Agriculture. In: F. L. Coleman [Ed.]. *Economic history of South Africa*. (pp. 37-85). Pretoria: Haum.
[34] Guelke, L. The white settlers, 1652-1780. In: R. Elphick and H. Giliomee (Eds.). *The shaping of South African society*, 1652-1820. (pp. 41-74), Cape Town, Maskew Miller Longman.

[35] Vrey, W. J. H. (1974). *'n Demografiese sosiologiese studie van die blanke bevolking in die Sentraal-Karoo [In Afrikaans]*. Bloemfontein: Institute for Social and Economic Research, University of the Free State.

[36] Giliomee, H. and Mbenga, B. (2007). *The new history of South Africa*. Cape Town: Tafelberg.

[37] Tamarkin, M. (1996). *Cecil Rhodes and the Cape Afrikaners: The imperial colossus and the colonial parish pump*. Johannesburg: Jonathan Ball Publishers.

[38] Nel, E. L. and Hill, T. R. (2008). Marginalisation and demographic change in the semi-arid Karoo, South Africa. *Journal of Arid Environments*, 72, 2264-2274.

[39] Fransen, H. (2006). *Old towns and villages of the Cape*. Johannesburg: Jonathan Ball Publishers.

[40] Myburgh, D. W. (1978). *The origin of towns in the Eastern Cape Midlands*, Contree, 4, 25-30.

[41] Christopher, A. J. (1984). *Colonial Africa*. London: Croom Helm.

[42] Blumenfeld, J. (1971). *The Economic Structure of the Cape Midlands and Karroo Region: A sectoral and spatial survey*. Grahamstown: Institute for Social and Economic Research, Rhodes University.

[43] Cook, G. (1971). *Towns of the Cape Midlands and Eastern Karoo*. Grahamstown: Institute for Social and Economic Research, Rhodes University.

[44] Le Clus, C. F. (1974). *Dienssentrums in die Sentraal-Karoo [In Afrikaans]*. Bloemfontein: Institute for Social and Economic Research, University of the Free State.

[45] Toerien, D. F. (2014). 'n Eeu van orde in sakeondernemings in dorpe van die Oos-Kaapse Karoo. (in Afrikaans). *Litnet* 11, 330-371.

[46] Toerien, D. F. (2012). Enterprise proportionalities in the tourism sector of South African towns. In: M. Kasimoglu, (Ed.), *Visions of global tourism industry: Creating and sustaining competitive strategies* (pp. 113-138) Rijeka: Intech.

[47] Toerien, D. F. and Seaman, M. T. (2012d). Regional order in the enterprise structures of selected Eastern Cape Karoo towns. *South African Geographical Journal* 94, 137-151.

[48] Toerien, D. F. and Seaman, M. T. (2014). Enterprise richness as an important characteristic of South African towns. *South African Journal of Science*, 110(11/12), Art. #2014-0018, 9 pages. http://dx.doi.org/10.1590/sajs.2014/20140018.

[49] *Wikipedia* (2015b). Occam's razor. Accessed on February 9, 1015 at http://en.wikipedia.org/wiki/Occam%27s_razor.
[50] Kennedy, P. (1993). *Preparing for the twenty-first century*. London: Fontana Press.
[51] Clark, G. (2007). *A farewell to alms: A brief economic history of the world*. Princeton, NJ: Princeton University Press.
[52] Clarke, K. R. and Warwick, R. M. (2001) *Change in marine communities: an approach to statistical analysis and interpretation*, 2nd ed. Plymouth: PRIMER-E.
[53] *Wikipedia* (2015c). Principal component analysis. Accessed on February 11, 2015 at http://en.wikipedia.org/wiki/Principal_component_analysis.
[54] Donaldson, R. and Marais, L. (2012). (Eds.) *Small town geographies in Africa: Experiences from South Africa and elsewhere*. New York, NY: Nova Science Publishers.
[55] Hoogendoorn, G. and Nel, E. (2012). Exploring small town development dynamics in rural South Africa's post-productivist landscapes. In: R. Donaldson and L. Marais (Eds.) *Small town geographies in Africa: Experiences from South Africa and elsewhere* (1st ed. pp. 21-34). New York, NY: Nova Science Publishers.
[56] Toerien, D. F. and Seaman, M. T. (2011). Ecology, water and enterprise development in selected rural South African towns. *Water SA* 37, 47-56.
[57] National Planning Commission. (2012). *National Development Plan 2030: Our future make it work*. Pretoria: The Presidency.
[58] Tilman, D. (1999). The ecological consequences of changes in biodiversity: A search for general principles. *Ecology* 80, 1455-1474.
[59] Dawes, R. M. (1979). The robust beauty of improper linear models in decision making. *American Psychologist*, 34, 571-582.
[60] Nyquist, S. and Lund, S. (2014). Shale revolution: Opportunity to jump-start economic growth. *Forbes Opinion*. Accessed on January 31, 2015 at: http://www.forbes.com/sites/realspin/2014/11/19/the-shale-revolution-is-an-opportunity-to-jump-start-economic-growth-in-u-s/.
[61] Agerton, M., Hartley, P., Medlock, B. M., and Temzelides, T. (2014). *Employment impacts of upstream oil and gas investment in the United States, Working Paper*. Houston, TX: James A. Baker III Institute For Public Policy, Rice University.
[62] Alford, E. (2015). *Eagle Ford Shale Drilling Rig Count*, February 13, 2015, Accessed on February 18, 2015 at http://eaglefordshale.com/drilling-rig-count/.

[63] Solomon, V. E. (1983) Transport. In: F. L. Coleman (Ed.) *Economic history of South Africa*. Pretoria: HAUM.

In: Arid and Semi-Arid Environments
Editor: Gabriel Ferguson

ISBN: 978-1-63483-383-7

Chapter 5

THIRTY YEARS OF ECOLOGICAL MONITORING IN ALGERIAN ARID RANGELANDS

Aziz Hirche[1,2,3], Mostefa Salamani[1,2], Boughani Abdelmadjid[1], Dalila Nedjraoui[1], Ratiba Hourizi[1], Youcef Oukil[1], Halima Slimani[1], Zina Khedache[1], Fahima Belala[1], Tahar Ait Hamouda[1], Nacer Adi[1] and Mohamed Grandi

[1]Laboratory of Vegetal Ecology, Faculty of Biological Sciences, University of Sciences and Technology Houari Boumediene (USTHB), Algiers, Algeria
[2]University of Algiers Youcef Benkhedda, Faculté centrale, Algiers, Algeria
[3]Scientific and Technical Research Center on Arid Regions, (CRSTRA), Biskra, Algeria

ABSTRACT

In 1978, several observatories, aiming ecological monitoring ,were settled in the arid high plains of South west Algeria. These observatories are now integrated in an international network called ROSELT (Long Term Ecological Monitoring Observatories Network) administered by the

OSS (Observatory of the Sahara and the Sahel). A study area included in one of these observatories was monitored on more than thirty years. It is part of Algerian steppe, the most widespread of the Maghreb. The results showed that this area has undergone profound changes.

These modifications concerned both physical and biological components. The parameters considered include vegetation, flora, and soil surface properties. It appears that land cover has changed and former vegetation dominated by *Stipa tenacissima, Lygeum spartum* and *Artemisia herba alba* has disappeared. New species, like *Atractyis serratuloides, Salsola vermiculata* and *Noaea mucronata*, rare in 1978 are now dominant, in relation with a dynamic of degradation. The largest decline is observed for *Stipa tenacissima* vegetation units constituting 2/3 of the landscape in 1978 and occupying just 1/10 in 2010. The vegetation cover has also significantly declined. In 1978, the average coverage was over 34%, whereas now it is less than 19% of the total. The biodiversity has also changed. The floristic richness has decreased. In 1978, 234 species were inventoried in the study area, for 134 in 2005 and only 94 in 2011. On the other side, the family composition and phytogeographical spectra have also changed, in the sense of an adaptation to xeric conditions. This can be explained by severe spells of drought combined by an exponential rise of livestock during the last 30 years. Finally, even if the last years were wetter and a slight improvement in vegetation cover was observed, there is no evidence for a desertification in reverse, as it is considered for the Sahel. On the contrary, we are still in a degradation process on the long term.

INTRODUCTION

The desertification of Algerian rangelands, mainly in arid and semi-arid lands, is a serious problem facing Algeria. The Algerian steppes are the most extensive rangeland of the North African countries. They occupy a pivotal position between the hilly and humid north, called the Tell, composing 5% of Algeria's surface, and the south formed by the Sahara covering 86% of the nation's territory. The demographic and the livestock increase, stepped up the pressure and contributed to the rangelands degradation. This trend seems to be observed not only in Algeria but in the whole Maghreb.

But, at an international level, several studies based on remote sensing data and techniques undertaken with the objective of monitoring land conditions found no evidence for extensive degradation in the Sahelian region (Helldén, 1984, 1991; Prince, Colstoun, Brown, and Kravitz, 1998; Rasmussen, Fog, and Madsen, 2001; Tucker and Nicholson, 1999). On the contrary, relatively recent studies of the Sahelian vegetation based on remote sensing data even emphasize "the greening-up" of the Sahel (Anyamba and Tucker, 2005; Herrmann, Anyamba, and Tucker, 2005). Some authors have extended the desertification study to the whole world and have concluded that greening-up is now, basically, a universal trend (Helldén, 2008; Helldén and Tottrup, 2008). Recently some recent studies have temperate the previous conclusions and have showed, in specific cases, a regressive trend (Dardel et al., 2014; Herrmann and Tappan, 2013). At the North African area, specifically at the Maghreb, the results are globally in contradiction with these results. The aim of this study is to provide recent data related to this problematic. It aims also to give some thoughts to this issue by evaluating qualitatively and quantitatively the phenomenon of land evolution in a representative area in the Algerian steppe.

The study area covers approximately 400 000 ha and the time horizon considered is about 30 years. In 1975, the Centre de Recherches sur les Ressources Biologiques et Terrestres (CRBT) carried out a major vegetation monitoring program in the south Oran steppe producing thematic maps of the whole area. Further, permanent observation sites have been set up to collect more accurate data.

The results will be confronted with those of neighboring countries. This allows providing a general overview for the North African countries and it would be interesting to carry out some comparisons with landscape changes within the Sahelian region.

Researchers of both the Sahelian and North African regions collaborate presently in the same major research project, ROSELT which constitutes an ideal framework to exchange scientific data and experiences.

1. Material and Methods

1.1. The Study Area

The study region, corresponding to El Biodh department (commune), covers an area of 374 400 hectares and is integrated in the Algeria's south-western High Plateaus. It is administered by the wilaya of Nâama (Figure 1) and the closest important city is Mecheria.

It is primarily occupied by the quaternary polygenic glacis with an average altitude of about 1000 m It includes saline depressions such as the Chott Chergui. The area is surrounded by chain of mountains such as the Antar djebel which culminates at 1700 m. This area is a part of Algerian ROSELT / OSS observatory.

The ROSELT (*Réseau d'Observatoires* de Surveillance Ecologique à Long Terme) is an international network aiming to monitor ecosystems on the long term implemented since 1994 by the observatory for the Sahara and the Sahel (OSS).

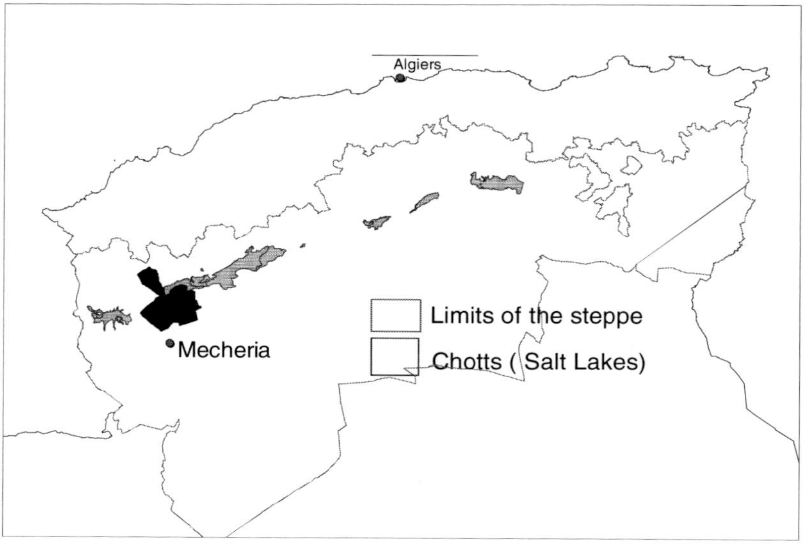

Figure 1. Location of the study area in the South west Algeria (Mecheria).

1.2. Climate

Climatic data was gathered from the National Meteorology Office and cover a long period from 1907 to 2013 (ONM, 2014). The climate is characterized by low annual rainfall with secular averages ranging from 264.21 mm in Mecheria the closest climatic station to 292.34 mm in El Bayadh. Since 1972, the rainfall has decreased and it is observed 229.15 mm in Mecheria to 270.43 mm in El Bayadh (a mountainous area surrounding the north-east). The distribution of rainfall is very irregular. The coefficient of variation reached 36.38% in Mecheria and 34.57% in El Bayadh. Winters are cold and harsh while summers are hot and dry. To illustrate the climatic context, the evolution of rainfall since the beginning of the century is shown in a graph (Figure 2). The rainfall evolution shows an alternation of dry and humid periods with no clear trends during approximately the first half of the century.

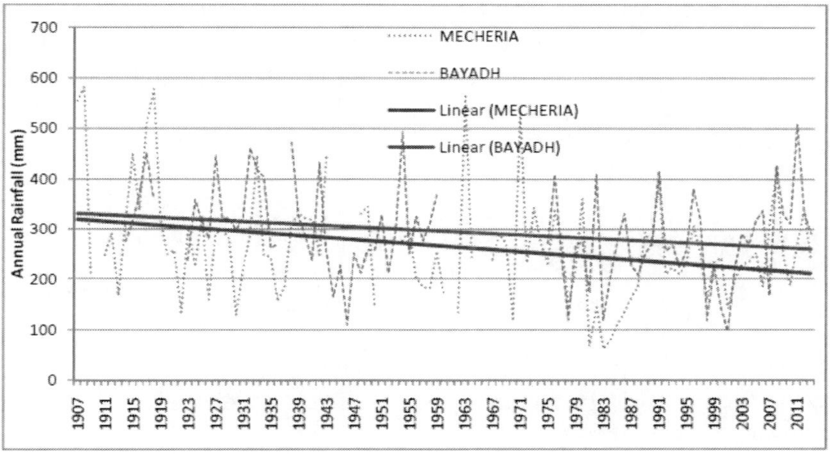

Figure 2. Annual Rainfall evolution since the beginning of the century in Mecheria and El Bayadh.

At the end of the second half, dry periods become notably more frequent and intense, especially during the 1981–1988 period (Hirche, Boughani, and Salamani, 2007) considered being the driest period of the last century. It is interesting to note that this drought period is partially the same as for the Sahel which has been affected by a notable large-scale drought during the 1983–1985 period after the first one that began in the late sixties. Since the end of the seventies, the mobile averages are almost always below the average rainfall.

The temperatures of respectively Mecheria and El Bayadh, show an average of minimum values for the coldest month (January) equal to 1.5°C and −1.8°C and an average of maximum values of the hottest month (July) equal to 35.1°C and 33.5°C. For all stations, the dry period (Bagnouls and Gaussen 1953) extends over 6 months and the bioclimate belongs to the medium-arid class with cold winters variant. Dust storms are rare at El Bayadh (3 days year - 1) but are more frequent in Mecheria (7.2 days year - 1). According to some authors, dust storms would be much more frequent since 1990, but until now, we do not have data related to this parameter before 1990 (Nouaceur, 2001). For the dust days (without storm), an average of 62 dust days per year was recorded in El Bayadh between 1989 and 2013, reaching 119 days per year between 2001 and 2004, considered as a drought period.

The same trend was observed in Mecheria, where 70 days were recorded between 1989 and 2013 and 165 days were observed in 2003 (Figure 3). The increase of dust days correspond globally to a decrease in rainfall.

In any event, if these results are proven over a longer series, they would be very alarming because they are related to the magnitude of soil denudation. The weak recovery of vegetation would then give free way to winds erosion over vast areas of the Algerian steppe.

1.3. Vegetation

The survey area was studied by different authors (Aidoud-Lounis, 1997; Aidoud, 1989; CRBT, 1978; Djebaili, 1990; Kadi-Hanifi Achour, 1998; Kadi-Hanifi, 2003) that described the vegetation dynamics in the steppic regions. A synthesis based upon these different studies is given in the following flowchart (Figure 4).

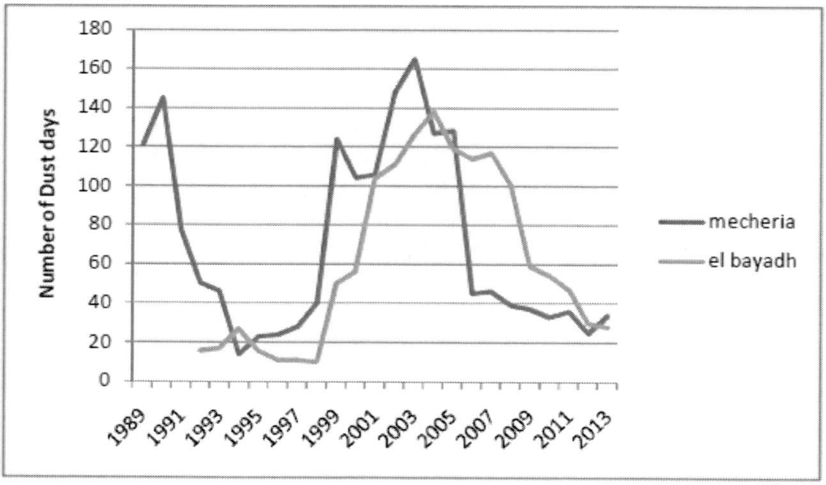

Figure 3. Evolution of the number of dust days in Mecheria and El Bayadh.

It appears that the Algerian steppe was initially colonized by forests and matorrals. Their gradual deterioration has led to the colonization of xerophytic vegetation consisting in a mixture of herbaceous and perennial plants.

The most emblematic is the herbaceous *Stipa tenacissima*, which colonized the majority of stony flat zones, corresponding to the geomorphologic units called "glacis."

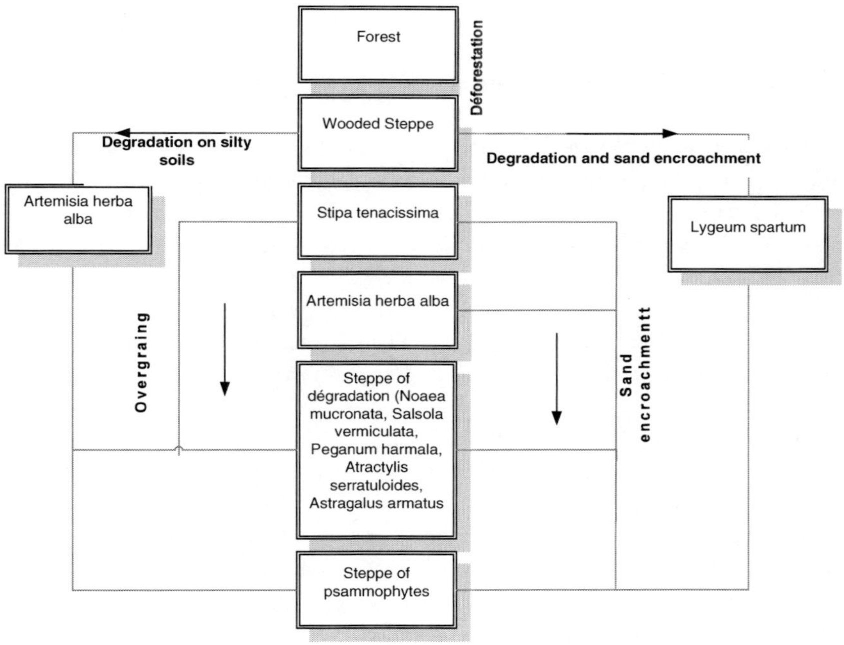

Figure 4. Main directions in the dynamic of vegetation in the study area.

Lygeum spartum, another herbaceous graminoid, appears when sand accumulates, often in depressions.

When degradation occurs, *Artemisia herba-alba*, a chamaephyt, appears on rich silt soils. These three species constituted the framework of the Algerian, and even the North African steppe.

Previous studies also showed that pressure on the environment, mainly by overgrazing, led these formations to disappear. They have been replaced by vegetation units or "formations" resulting from a dynamic degradation with less functional performances, such as *Atractylis serratuloides, Salsola vermiculata, Peganum harmala*, and *Astragalus armatus*. The conversion of arid grassland to shrublands due to overgrazing and/or drought is a well known phenomenon. Allington and Valone (2010) reviewed the main concerned models and concluded that "current conceptual models of desertification predict that arid grasslands exist in one of two alternate stable states over timescales relevant to management: perennial grasslands or desertified shrubland." When sand accumulation becomes prominent, psammophytes such as *Thymelaea microphylla* settle.

1.4. Data Collection and Processing

The current work was divided into several stages. The first was to establish land cover maps at different dates, 1978, 2005, and 2011. The last two maps were realized through a photo-interpretation of Landsat images (TM Landsat) to have a product comparable in its realization to the old land cover maps (1978).

In each map, different units of vegetations were considered and were described by their physiognomic types. Theses ones are defined by the first (up to three) dominant species. Each vegetation unit is also informed by the class cover vegetation.

On the basis of these land cover maps with a temporary key legend, a sampling work was carried at each period. It was a mixture sampling in the sense of Gounot (1969).

In each vegetation unit, were carried out several phytoecological releves, describing vegetation and flora. They were accompanied by information related to the state of the ground surface (coarse elements, pebbles, gravel, sand.). Finally a definitive key legend is established.

2. Land Cover Maps

The land cover maps corresponding to each period are illustrated in the figure 5 a, b, c. The Table 1 reports the main vegetation units existing in the study area, with their areas. It shows a clear contrast between 1978 and 2005. Usually, the steppe vegetation was dominated by three species. The first ones are graminaceous; the alfa (*Stipa tenacissima*) and the "sennagh" (local vernacular name) or esparto (*Lygeum spartum*). The third one, the "chich" or wormwood is chamephytic. The green, grey and the orange colors dominate the land cover map of 1978. The green color corresponds to *Stipa tenacissima*, a keystone ecological species. The orange color represent the *Artemisia herba alba* a species appreciated by the cattle. The grey color corresponds to *Lygeum spartum* a psammophyt plant which has a large capacity of recovery. Since

1978, these "pseudo" climactic vegetation units, with significant vegetation covers, generally superior to 30% regressed.

Stipa tenacissma and *Lygeum spartum* vegetation units lost 8/10 of their original area, and it reaches even 9/10 for the *Artemisia herba alba* ones. The average vegetation cover has decreased and reached only 17% on 2005 and 19% on 2011.

Table 1. Evolution of the areas of the vegetation units

Formation	1978		2005		2011	
	Ha	%	Ha	%	Ha	%
Matorral	3908.74	0.56	6849.80	0.98	6849.80	0.98
Stipa tenacissima	313445.08	44.76	46679.52	6.67	11496.93	1.64
Artemisia herb alba	8441.51	1.21	805.58	0.12		
Lygeum spartum	348097.22	49.70	55 607. 60	7.94	111960.74	15.99
Steppe of degradation		0.00	489168.97	69.85	482412.60	68.88
Salty depressions	21030.00	3.00	29609.03	4.23	27625.84	3.94
Crops	419.21	0.06	8880.67	1.27	8366.01	1.19
Psammophyts	5000.94	0.71	59851.71	8.55	48741.06	6.96
Urban			2889.25	0.41	2889.25	0.41

In the same time, new vegetation units called "steppe of degradation," issued from a dynamic degradation have appeared (Figure 4). These units are dominated by species like *Atractylis serratuloides, Peganum harmala and Salsola vermiculata.*

The Table 1 also shows that the difference between 2005 and 2011 is not as important as between 1978 and 2005. No new vegetation unit does appear in the meantime.

It is just observed a reshuffle in terms of area. The *Stipa tenacissima* vegetation units still decrease and lost the three quarters of their original area. *Artemisia herba alba* finally disappeared completely in 2011. Only the Lygeum spartum amongst the original vegetation, after a drastic fall of its area between 1978 and 2005, increased anew between 2005 and 2011, twice in terms of area. It is interesting to note that, in this last period, the steppe of degradation does not increase more and the vegetal canopy even slightly increases (2005-2011).

3. Evolution of the Biodiversity

Species richness was determined by counting the number of species on the basis of all floristic lists. Species richness does not constitute a reliable indicator if it is not compared and related to rainfall over a long time period.

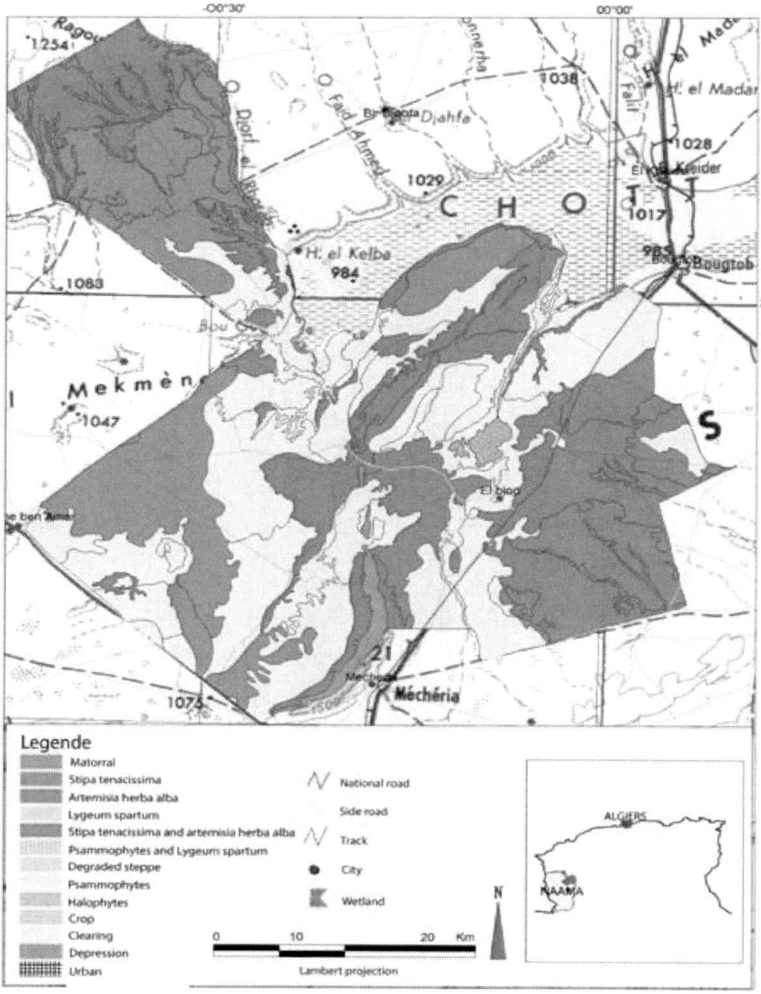

Figure 5a. Land cover map of El Biodh (1978).

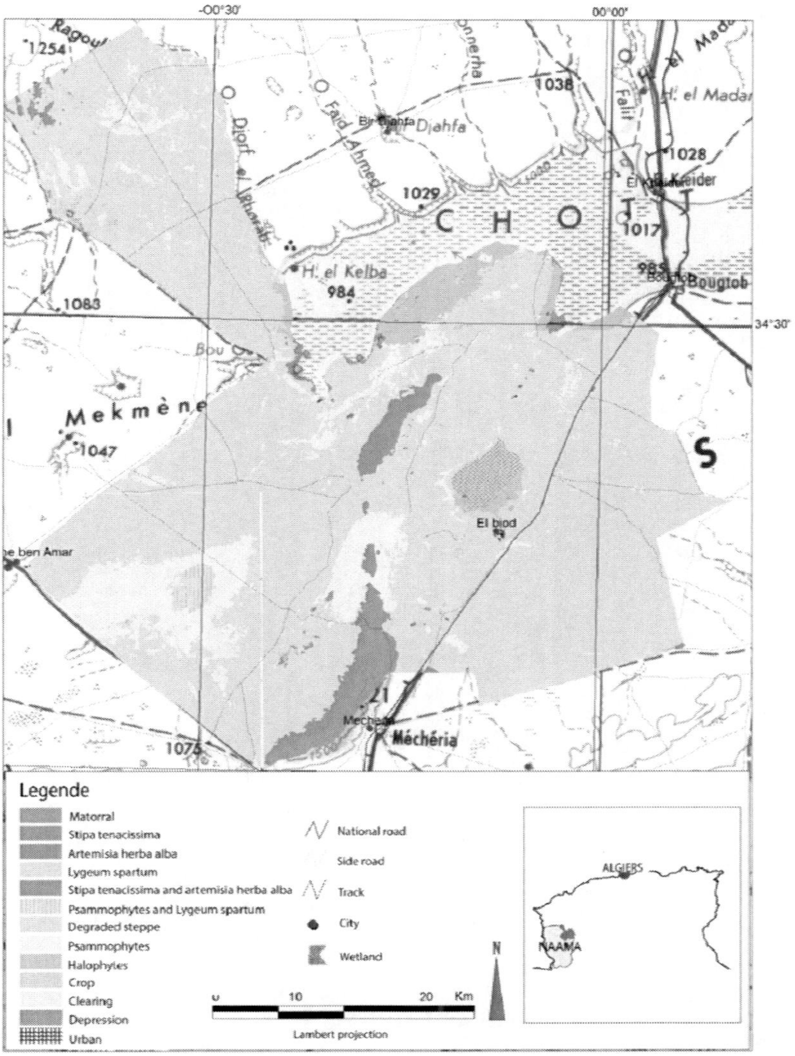

Figure 5b. Land cover map of El Biodh (2005).

In our study area, integrated in the ROSELT / OSS network, the established trends showed a qualitative and quantitative decline of vegetation which is also expressed by a decrease in species richness.

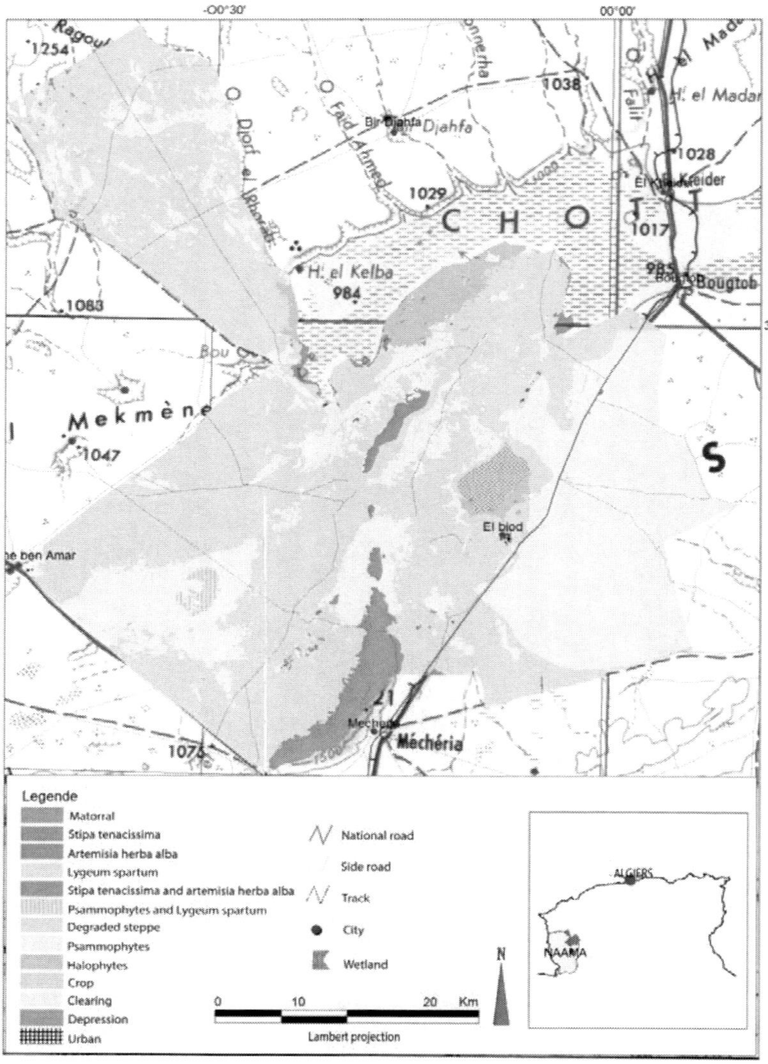

Figure 5c. Land cover map of El Biodh (2011).

The total number of species shows a marked decrease between periods considered. Thus, one passes successively of 234 species in 1978 to 134 species in 2005 and 94 species in 2011. The number of species disappearing is more important for annuals than for perennial and exceeds the half of initial stock.

Curiously, the "therophytization," a classical trend which occurs when the desertification is more intense and which means an increase of annuals is not observed in the present study. It could be related to a decrease of seed bank in the soil. Although little studied in North Africa, the phenomenon of loss of stock in seeds in the soil is a parameter that is commonly given in the Sahel. (Hiernaux, Diarra, and de Leeuw, 1991; Hiernaux et al., 2009) and (Grouzis, 1992) showed that overgrazing tended to decrease the stock seed.

Table 2. Evolution of vegetation covers of the vegetation units

Formation	1978			2005			2011		
	Ha	VC %	VCP %	Ha	VC %	VCP %	Ha	VC %	VCP %
Matorral	3908.74			6849.80			6849.80		
Stipa tenacissima	313445.08	36		46679.52	20		11496.93	20	
Artemisia herb alba	8441.51	28		805.58	12				
Lygeum spartum	348097.22	34		55 607. 60	19		111960.74	21	
Steppe of degradation				489168.97	16		482412.60	19	
Salty depressions	21030.00	38		29609.03	26		27625.84	28	
Crops	419.21			8880.67			8366.01	15	
Psammophyts	5000.94	18		59851.71	23		48741.06		
		30.8	34.62	2889.25	19.33	17.10	2889.25	20.6	18.92

VC: Vegetation covers. VCP: Vegetation cover weighted by area.

Table 3. Comparison in evolution of perennial and annual species through the three periods

Period	1978		2005		2011		Evolution of Total (2011-78)	
	Nbr of species	%	Nbr of species	%	Nbr of species	%	Nbr of species	% diminution
Perennial	129	55.12	76	56.72	63	67.021	66	-51.11
Annual	105	44.87	58	43.28	31	32.98	74	-70.47
Total	234	100	134	100	94	100	140	-59.82

The edaphic aridity could be a second explanation since soils have undergone significant deflation and the rising of temperatures have led to a negative impact on vegetal production. It is noted significant changes in the list of observed species. Some of them are found only in the first period as: *Brachypodium distachyum, Sedum sediforme, Ziziphora hispanica and*

Xeranthemum inapertum, and to a lesser extent *Androsace maxima, Bromus squarrosus and Micropus bomycinus*. These species would be indicator of a wetter climate (Aidoud-Lounis, 1997; Aïdoud, Le Floc'h, and Le Houérou, 2006; Aidoud, Slimani, and Rozé, 2010; Le Houérou, 1969). Their absence in 2005 and 2011 would testify also on the degradation of soils (Slimani et al., 2010). These species have been replaced by others indicating degradation as *Atractylis cancellata, Atractylis flava* and *Xanthium spinosum. Salsola kali* appeared only in the third period.

3.1. Evolution of Perennial and Annual Species

Monitoring the perennial species can be a better indicator than the global list of species. Indeed, perennial species are less correlated to climatic hazards as annuals. The results show a significant decrease in the number of perennial species by referring to the first period (1978). The number of perennial species shows a decrease of 53 species during the second period of observation (2005) and a cumulated number of 66 species till the third period (2011). For the total, nearly the two third of initial species has momentarily disappeared. The changes in species composition would affect ecosystem function, but there is several hypothesis about the ecosystems respond to changes in species richness (Naeem et al., 2002). Nevertheless, it appears that keystone hypothesis is particularly relevant in this current case. A rapid decline of ecosystem functions was observed with the loss of crucial species and diversity is consequently reduced below its natural level (Walker, 1992).

The disappearance of a keystone species of the original dominant steppe, namely the alfa has deep impact on the ecosystem (Kadi-Hanifi Achour, 1998; Kadi-Hanifi, 2003; Slimani and Aidoud, 2004; Slimani, Aidoud, and Rozé, 2010). Degradation units like *Salsola vermiculata, Thymelaea microphylla and Peganum harmala* naturally show a lower efficiency in the functioning of ecosystems. They show also a lower phytomass and richness than alfa vegetation even if ecological conditions are favorable (rainfall).

3.2. Evolution of the Systematic Composition

The decrease in the overall number of species through the 3 periods has resulted in a change in the distribution of the types of families throughout the monitoring period. The evaluation of this change was made taking into consideration the number of species in each family (Table 4).

Figure 6 shows that there is a decrease in the number of species in several families. This reduction does not have the same importance for all of them and allows to rank them in order of importance.

The decline rate is important for three families: Poacea, Brassicaceae, Caryophyllaceae, Apiaceae and Chenopodiaceae (Amaranthaceae), especially during the third period. The diminishing of Asteraceae is real but comparatively to the previous ones, this family seems more stable in its size. It seems more "equipped" to resist to changing conditions and ecological anthropogenic pressures in these degraded environments.

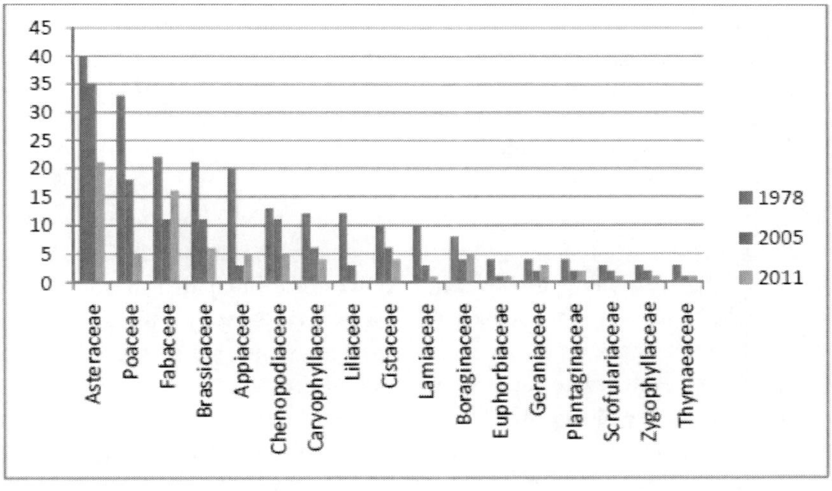

Figure 6. Evolution of the number of species for the main families through the three periods.

3.3. Evolution of Biogeographical Spectra

The study of biogeographical elements is interesting since we can determine through their evolution, the impact of the aridity on the species distribution. The Figure 7 represents the evolution of biogeographical spectra in the study area during the 3 periods. It appears that the Mediterranean species are decreasing, since Saharo -Sindian and irano-Touranian, more xeric elements, increased.

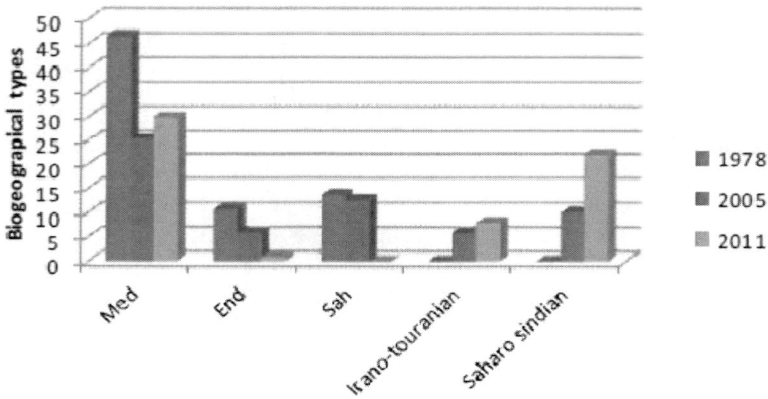

Figure 7. Evolution of some biogeographical elements through the three periods.

4. Socio-Economic Features

Mecheria is a pastoral area. The rangelands which constitute the main support to the local economy were recently degraded, especially since the eighties. At the beginning of the century, livestock fluctuated in correlation with rainfall. When droughts occurred, livestock decreased and the cattle were decimated. When the rainfall becomes more consistent, the livestock become more important. A dynamic equilibrium was realized.

After the independence of the country, the improvement of prophylaxis and the using of complementary aliments allowed to a relative independence of climatic conditions for the livestock. So, its size increased exponentially whatever the annual rainfall (Hirche, Salamani, Abdellaoui, Benhouhou, and Valderrama, 2011). Meanwhile, the rainfall in this period is the lowest of the century.

An imbalance between the fodder production and the alimentary needs of the cattle occurred and becomes increasingly important. The period 1982-1988, which was among the driest of the century, showed a very strong degradation of the rangelands and led to the progressive disappearance of *Stipa tenacissima and Artemisia herba alba.* These species were certainly weakened by drought, but their disappearance is mainly due to overgrazing. Indeed, the exclosure established in the study area clearly showed that the alpha protected survive and does not disappear.

5. Discussion and Interpretation

The study of indicators produced from the land cover maps performed on three different dates in the El Biodh ROSELT observatory, highlighted converging trends between 1978 and 2011, showing a very important degradation of vegetation. The dominant formations (vegetation units) such as *Stipa tenacissima* and to a lesser extent *Artemisia herba alba*, a keystone species, disappear relatively quickly and are replaced by new vegetation units resulting from a dynamic degradation like *Atractylis serratuloides* or *Salsola vermiculata*. The disappearance of these vegetation units is quite fast since it spanned only about two decades while these formations have existed for at least a century and probably much older This implies a profound transformation of landscape of the region materializing mainly by fragmentation as shown by (Hirche, Salamani, Abdellaoui, Benhouhou, and Valderrama, 2011). Socio-economic changes, clearing, overgrazing and timber harvesting have led to a significant fall for all the indicators (vegetation cover, phytomass ...).

By cons, between 2005 and 2011, no significant regressive changes were observed and were visible through the indicator set. In contrast, for some of them, the trend has reversed, like cover vegetation which shows a very slight biological recovery and a relative rapid recolonization of *Lygem spartum*.

The main degradation factors, namely the demographic and livestock increase which causes increased pressure on rangelands by overgrazing are always present and don't herald any improvement in the long term. The recent greening would be due to favorable rains in recent years and the slight decrease in livestock in Algeria due to the scarcity of foods, including those imported. Our results are corroborated by similar ones led in Tunisa by Hanafi and Jauffret (2008) and the ROSELT synthesis on Ecology for North and West Africa (2013). Benbrahim et al. (2004) and Mahyou, et al. (2010) showed the same trend in Morocco, in their arid rangelands which represent the western extent of our study area.

The whole North Africa rangelands shows a drastic changes by the disappearing of original dominating species, mainly herbaceous and their replacing by new chamaephyte vegetation with lesser cover and phytomass.

The degradation rate is much more rapid than the restoration speed. Even, this decade was more or less wet, no significant improvement was observed notwithstanding slight increasing of vegetation cover and a decrease of sand cover. Degradation phases were so intense that today, despite a return of the rains, annual struggling to germinate and return to their original abundance. In this analysis, we must remain vigilant to talk about a total disappearance of a species because only continuous and comprehensive observations can confirm these preliminary conclusions, especially when it comes to annual or short-lived species that are more influenced by inter-annual variations in rainfall. These findings seem to contradict many "remote-sensed" studies around the world, especially in the Sahel region, showing instead a greening up and correspondingly the absence or the stopping of desertification phenomenon. Indeed, based on the images satellite, and studying the evolution of NDVI provided by GIMMS images over the years, Hellden (1984, 1991); Prince et al. (1998), Tucker and Nicholson (1999); Rasmussen et al. (2001), Herrman and Tucker (2005) and (Olsson, Eklundh, and Ardö, 2005) found no evidence of extensive degradation of the Sahel. Already, Thomas, in 1993, goes further and published a book in provocative title "Sandstorm in a teacup, taunting and mocking proponents of desertification. Some authors extend their findings, again from satellite image observations, to the whole world (Hellden 2008; Hellden and Tottrup 2008).

We have used the same images and the same methodological approach, and the "remote sensed" results are conforming to the results to the different authors cited above (Hirche et al., 2010) and indicate a rise in NDVI values. The increasing of NDVI refers to the rainfall increase of the last years which induced a vegetation growth and a cover vegetation rise. However, the soils are wetter and no longer subject to wind erosion, inducing a sand coverage decreases. These parameters influence the spectral response and contribute to decrease the value of NDVI. Therefore, the trend shown in a decrease of NDVI spectral response is related to several factors, not just an increase in vegetation cover and phytomass. We have shown through this survey that vegetation shows instead a regressive trend although the desertification seems to have slowed in recent years. Even in the Sahel, recent studies have temperated the previous conclusions of a greening-up and have showed in specific cases a regressive trend (Dardel et al., 2014; Herrmann and Tappan, 2013). It is interesting to observe that these recent works address perennial vegetation that is the exception rather than the rule in the Sahel.

It seems crucial when we consider and interpret the desertification to a better understanding to distinguish between dominant annual vegetation such as at the Sahel and dominant perennial vegetation such as the North Africa.

Yet recent famines in the Sahel do not tend to sit the thesis of a biological recovery ("Greening up"). Some author like Ozer et al. (2010) refuted the greening- up of the Sahel.

All this discussion, shows that the field work associated to the regular monitoring of rangelands remain indispensable for understanding the functioning of Ecosystems.

Conclusion

This study was led at El Biodh, a representative region which summarizes the main trends of Algerian rangelands and even North African rangelands.

All the results through the study of vegetation and flora converge and show a degradation trend. But, significant difference exists between the first (1978-2005) and the second period (2005-2011).

The main changes occurred during the first period which is close to 30 years. The second period doesn't show an obvious regressive trend. On the contrary, the cover vegetation is slowly increasing, but our conclusions have to been moderated because the period is too short and the two periods are not completely comparable. In addition, despite the increase in vegetation cover, vegetation dynamics remains regressive.

The recent greening-up does not mean a desertification in reverse. In fact, desertification is process on the long term, and recent changes seem to be more a fluctuation than a trend. Only a regular monitoring of the vegetation as major component of the ecosystem, allows a real understanding of desertification process.

References

Aidoud-Lounis, F. (1997). Le complexe à alfa-Armoise -sparte (Stipa tenacissima L., Artemisia herba-alba Asso, Lygeum spartum L.) des steppes arides d'Algérie. *Structure et dynamique des communautés végétales* (198 pp.). Doctoral thesis, University of Aix-Marseille III, France.

Aidoud, A. (1989). Contribution à l'étude des écosystèmes steppiques pâturés des Hautes Plaines Algéro-Oranaises. Fonctionnement, évaluation et évolution des ressources végétales. *Doct. Thesis; Uni. Sci. Techn. H. Boumediène. Alger, 253p. + Ann.*

Aïdoud, A., Le Floc'h, É. and Le Houérou, H. N. (2006). Les steppes arides du nord de l'Afrique. *Science et changements planétaires/Sécheresse*, 17(1), 19-30.

Aidoud, A., Slimani, H. and Rozé, F. (2010). La surveillance à long terme des écosystèmes arides méditerranéens: quels enseignements pour la restauration? Cas d'une steppe d'Alfa (Stipa tenacissima L.) en Algérie. *Revue internationale d'écologie méditerranéenne International Journal of Mediterranean Ecology.*

Allington, G. and Valone, T. (2010). Reversal of desertification: the role of physical and chemical soil properties. *Journal of Arid Environments*, 74 (8), 973-977.

Anyamba, A. and Tucker, C. (2005). Analysis of Sahelian vegetation dynamics using NOAA-AVHRR NDVI data from 1981–2003. *Journal of Arid Environments*, 63(3), 596-614.

Benbrahim, K. F., Ismaili, M., Benbrahim, S. F., and Tribak, A. (2004). Problèmes de dégradation de l'environnement par la désertification et la déforestation: impact du phénomène au Maroc. *Science et changements planétaires/Sécheresse*, 15(4), 307-320.

CRBT. (1978). Rapport phytoécologique et pastoral sur les hautes plaines steppiques de la wilaya de Saida. *Algiers. ONRS.*

Dardel, C., Kergoat, L., Hiernaux, P., Mougin, E., Grippa, M., and Tucker, C. (2014). Re-greening Sahel: 30years of remote sensing data and field observations (Mali, Niger). *Remote Sensing of Environment*, 140, 350-364.

Djebaili, S. (1990). Syntaxonomie des groupements préforestiers et steppiques de l'Algérie aride. *Ecologia mediterranea*, 16, 231-244.

Gounot, M. (1969). *Méthodes quantitatives d'étude de la végétation*: Editorial Masson, París.

Grouzis, M. (1992). Germination et établissement des plantes annuelles sahéliennes. L'aridité: une contrainte au développement: caractérisation, réponses biologiques, stratégies des sociétés, E. Le Floc'h, M. Grouzis, A. Cornet, J.-C. Bille, *Editeurs scientifiques, ORSTOM*, 267.

Helldén, U. (1984). Drought impact monitoring. A remote sensing study of desertification in Kordofan, Sudan. *Rapporter och Notiser-Lunds Universitets Naturgeografiska Institution* (Sweden).

Helldén, U. (1991). Desertification: time for an assessment? *Ambio*, 372-383.

Helldén, U. (2008). A coupled human–environment model for desertification simulation and impact studies. *Global and Planetary Change*, 64(3), 158-168.

Helldén, U. and Tottrup, C. (2008). Regional desertification: A global synthesis. *Global and Planetary Change*, 64(3), 169-176.

Herrmann, S. M., Anyamba, A. and Tucker, C. J. (2005). Recent trends in vegetation dynamics in the African Sahel and their relationship to climate. *Global Environmental Change*, 15(4), 394-404.

Herrmann, S. M. and Tappan, G. G. (2013). Vegetation impoverishment despite greening: A case study from central Senegal. *Journal of Arid Environments*, 90, 55-66.

Hiernaux, P., Diarra, L. and de Leeuw, P. (1991). Dynamique de la vegetation sahelienne apres secheresse. Un bilan du suivi des sites pastoraux du Gourma en 1990. *ILCA Mali Document de Travail (ILCA). no. 001/91.*

Hiernaux, P., Mougin, E., Diarra, L., Soumaguel, N., Lavenu, F., Tracol, Y., and Diawara, M. (2009). Sahelian rangeland response to changes in rainfall over two decades in the Gourma region, Mali. *Journal of Hydrology*, 375(1), 114-127.

Hirche, A., Boughani, A. and Salamani, M. (2007). Évolution de la pluviosité annuelle dans quelques stations arides algériennes. *Science et changements planétaires/Sécheresse*, 18(4), 314-320.

Hirche, A. (2010). Contribution à l'évaluation de l'apport de la télédétection spatiale dans la dynamique des écosystèmes en zones arides: Cas du Sud oranais. *Doctorate. Thesis. University. of Sciences and Technology Houari Boumediene. 330 p+ ill. Algiers.*

Hirche, A., Salamani, M., Abdellaoui, A., Benhouhou, S., and Valderrama, J. (2011). Landscape changes of desertification in arid areas: the case of south-west Algeria. *Environ. Monit. Assess.*, 179(1-4), 403-420. doi: 10. 1007/s10661-010-1744-5.

Kadi-Hanifi Achour, H. (1998). L'alfa en Algérie. Syntaxonomie, relation milieu-végétation, Dynamiqueet perspectives d'avenir. *Thèse. Doct. Es science. Univ. Sci. Tech. H. Boumediene. Alger. 270 p.*

Kadi-Hanifi, H. (2003). Diversité biologique et phytogéographique des formations à Stipa tenacissima L. de l 'Algérie. *Science et changements planétaires/Sécheresse*, 14(3), 169-179.

Le Houérou, H. N. (1969). *La Végétation de la Tunisie steppique: structure, écologie, sociologie, répartition, évolution, utilisation, biomasse, productivité (avec références aux végétariens analogues d'Algérie de Libye et du Maroc).* Thèse. Doc. D'état.

Mahyou, H., Tychon, B., Balaghi, R., Mimouni, J., and Paul, R. (2010). Désertification des parcours arides au Maroc. *Tropicultura*, 28(2), 107-114.

Nouaceur, Z. (2001). L'évolution des lithométéores en Mauritanie confirme une plus grande efficacité de l'action éolienne. *Cybergéo*.

Olsson, L., Eklundh, L. and Ardö, J. (2005). A recent greening of the Sahel—trends, patterns and potential causes. *Journal of Arid Environments*, 63(3), 556-566.

Ozer, P., Hountondji, Y., Niang, A. J., Karimoune, S., Laminou Manzo, O., and Salmon, M. (2010). Désertification au Sahel: historique et perspectives. *Bulletin de la Société Géographique de Liège*, 54, 69-84.

Prince, S. D., Colstoun, D., Brown, E., and Kravitz, L. (1998). Evidence from rain-use efficiencies does not indicate extensive Sahelian desertification. *Global Change Biology*, 4(4), 359-374.

Rasmussen, K., Fog, B. and Madsen, J. E. (2001). Desertification in reverse? Observations from northern Burkina Faso. *Global Environmental Change*, 11(4), 271-282.

Slimani, H. and Aidoud, A. (2004). Desertification in the Maghreb: a case study of an Algerian high-plain steppe *Environmental Challenges in the Mediterranean 2000–2050* (pp. 93-108): Springer.

Slimani, H., Aidoud, A. and Rozé, F. (2010). 30 Years of protection and monitoring of a steppic rangeland undergoing desertification. *Journal of Arid Environments*, 74(6), 685-691.

Tucker, C. J. and Nicholson, S. E. (1999). Variations in the size of the Sahara Desert from 1980 to 1997. *Ambio*, 587-591.

INDEX

E

F

G

H

I

J

K

L

M

N

O

P

Q

R

S

X

Y